THE SHORT-RANGE ANTI-GRAVITATIONAL FORCE AND THE HIERARCHICALLY STRATIFIED SPACE-TIME GEOMETRY IN TWELVE DIMENSIONS

THE SHORT-RANGE ANTI-GRAVITATIONAL FORCE AND THE HIERARCHICALLY STRATIFIED SPACE-TIME GEOMETRY IN TWELVE DIMENSIONS

CHRISTINA ANNE KNIGHT

To order additional copies of this book, contact:
Xlibris LLC
1-888-795-4274
www.Xlibris.com
Orders@Xlibris.com
83529

"Imagination is more important than knowledge."

—Einstein

"The only real valuable thing is intuition."

—Einstein

Preface

There is increasing evidence that much about the universe that scientists believe may be wrong. There is also increasing awareness that these same scientists are missing something fundamental as they carry on in their quest for a theory of everything and a theory that unites general relativity with quantum mechanics. Nevertheless, it is quite possible that we live in a wondrous time in which many of the important questions may finally be answered. However, this will probably require revolutionary thinking and many widely accepted ideas will become casualties in what will be a revolutionary overhaul of physics and cosmology. In this work, I propose a new theory that suggests that the space-time geometry possesses a complex hierarchical structure that comprises twelve dimensions (nine space dimensions and three time). Furthermore, this structure is divided into three strata, each of which has its own four-dimensional structure and stratum-specific parameters (with variations in the gravitational constant G, the speed of light c, and the Planck constant).

Among the issues confronting cosmologists are questions concerning the apparent *fine-tuning* of the parameters in our present universe (as well as the low entropy state that existed at the beginning of it). My thesis is that such fine-tuning (as well as the initial low entropy state) can be explained by means of an evolutionary cyclic model that restricts the role of the 2nd Law of thermodynamics in cosmic evolution. In addition, all of the stratum-dependent variations in the parameters that exist during the present cosmic cycle are the product of a long, transcyclic, evolutionary history, evolving via a process that I call *parametric mutagenesis*. Moreover, there is also a stratum-dependent variation in the nature and range of the fundamental forces that operate within the tri-stratum, space-time structure.

If space-time is quantized (which suggests that singularities cannot exist), there must also exist a force that counteracts the gravitational force at Planck length distances (or shorter). The existence of a short-range antigravitational force may provide the means by which space-time quantization is maintained. Such a force may also be important in explaining the cause of the initial big bang expansion. The differences in the range and the nature of this force and those of the gravitational force result in an unstable relationship. It is this instability that is responsible for producing the cosmic thermodynamic gradient, and through the process of cyclic evolution, the universe becomes increasingly more efficient at reducing this cosmic gradient (up to a point as you will see).

It is hoped that the reader will find these ideas, as well as others mentioned in this work, thought provoking. The text is presented in a manner that should be intelligible to an intelligent layperson who has some basic knowledge of physics and cosmology. In addition, this work is not comprehensive in scope but serves as a vehicle to introduce new ideas. Some omissions were necessary for the sake of brevity; and some, because the author did not consider them relevant to the main issues discussed. Certain topics like that concerning the Higgs boson are not mentioned because the author does not believe that the Higgs field exists (in fact, I predict that the Large Hadron Collider at CERN will be unsuccessful in its attempt to find it). Nevertheless, it is hoped that some important finds will be made in the experiments performed at the Large Hadron Collider. Perhaps, some of the ideas presented by this author will be confirmed. In any case, if the basic premise of this work is correct, it will be necessary to determine the values for the parameters operating in each of the strata. Once this is accomplished, the answers to many questions may be in reach.

Christina Anne Knight

Newport News, VA
June 19, 2010

Many people have chosen their personal heroes from the ranks of gifted athletes and popular celebrities. This is, of course, a reflection of individual interest and taste. For me, the real heroes are those individuals who have made profound contributions to our knowledge about the world and the universe in which we live. However, the esteem in which I hold my own personal heroes, Charles Darwin and Albert Einstein, stems as much from the biographical similarity of their inspiring meteoric rise from relative obscurity, as it is their accomplishments in their respective fields. Although the concept of biological evolution was not new when he arrived on the scene, Darwin was able to construct a convincing theory that could adequately explain the wide diversity of life and its complexity. He accomplished this through the introduction of the concept of natural selection as the primary mechanism driving evolutionary processes.

Of course, the work of Gregor Mendel introduced an additional mechanism with his early work on genetics, which provided an ingredient absent from Darwin's theory. It remains to be seen whether the current neo-Darwinian synthesis will undergo further modification if a superior causal explanation can be discovered for genetic mutation. The *Bénard cell experiment*, which demonstrates how order may emerge spontaneously from disorder upon the attainment of a critical threshold, may be instructive, and conclusions derived from it may have broader applications. If this is so, it may very well be that genetic mutations may emerge spontaneously within a genetic line when critical thresholds are attained (owing to thermodynamic pressures) through the copious reproduction of interacting genes over the course of many generations. It may be that the barriers (for example, geographic isolation) that introduce genetic isolation may only increase the rapidity with which these thresholds are attained. My own intuition is that beneficial genetic mutations do not emerge merely as random errors in genetic copying but emerge inevitably when the optimum *intergenetic* and thermodynamic conditions permit them to do so. Regardless of whether further modification to existing theory is necessary, there is no question that biological evolution is a scientific fact supported by overwhelming evidence. It also appears that evolutionary processes have a broader, universal application and are involved in some form with all systems that constitute the hierarchy of nature. This includes physical systems and the more abstract systems such as language and culture.

The complexity of the universe as evidenced by its apparently *fine-tuned* parameters, as well as the rapidity with which relatively complex structures emerged early in cosmic development, requires an explanation. It is somewhat ironic that the field of cosmology is within a position not altogether different from that of biology before the advent of the ideas provided by Darwin and Mendel. Before Darwin, the favored explanation for biological complexity was the benevolent machinations of a transcendent deity (an external agency). In a similar vein, many cosmologists today propose the existence of a multiverse (a different kind of transcendent agency involved here—probability) to explain the apparent fine-tuning observed within our universe. According to the multiverse hypothesis, there is perhaps an infinite number of other universes coexisting externally to our own, and each is governed by perhaps an infinite range of varying parameters. The reasoning here is that in the midst of such a large number of universes, the probability that at least one of them may have the parameters that are found in our own universe is reasonably high.

It is acknowledged that none of these universes interact,[1] nor is there likely to be any means to detect the presence of these empirically. This suggests that any causal explanation of a given universe's parameters should be sought within its own existential context. This means that the apparent fine-tuning is *autocausally* dependent on the individually specific and inherent properties that drive the evolutionary process. Rather than being the product of a lucky throw of the *cosmic dice*, it seems more likely that the currently existing parameters have an extensive evolutionary (hereditary?) history that can only be accommodated within the context of an *evolutionary, cyclic model*. If this is true, then it must be determined which mechanisms, including what fundamental properties of the space-time geometry, are responsible for cosmic evolution. To accomplish this, it may require the discovery of a means by which the second law of thermodynamics can be circumvented, modified, or recontextualized (in spite of Eddington's admonition

---

[1]   It has been proposed that the relative weakness of gravity in our universe is accounted for if the large proportion of it is leaked into parallel universes through the curled up extra dimensions, the existence of which is predicted by string theory, but for the most part these universes do not interact.

that any theory that is inconsistent with the second law is doomed to failure).

That some form of selection process (perhaps *thermodynamically driven*) may be involved in cyclic evolution is possible, although it may not take the same easily recognizable form as that theorized for biological evolution. If the universe and its evolution is self-contextual and self-induced, this eliminates the notion of competing variants (on the same systemic level) vying for selection, as is the case in biological evolution. Therefore, if cosmological selection is to *work*, we must find an alternative means by which it may operate that does not involve competition between variant universes (*phenotypes*) and parameters (*genetic codes*). There has been one notable attempt to incorporate natural selection into a model of cosmic evolution by the physicist Lee Smolin (a leading proponent of loop quantum gravity). In his model, *cosmological natural selection* favors universes that maximize black hole formation. From these black holes, new universes spring forth, and of these offspring universes, those which continue to maximize black hole formation are selected. Those universes that do not maximize black hole formation may become the latest victims of the selection process. Dr. Smolin is to be applauded for at least making an attempt to explain the apparent fine-tuning of parameters within an evolutionary context and adopting an evolutionary mechanism rather than accepting the status quo.

It certainly would be preferable, however, if an application could be found without recourse to the unparsimonious multiverse notion in any form. In addition, it is probable that black hole formation is an inevitable process that has a fundamental role in the *cyclic evolutionary process* that has not yet been recognized. An *evolutionary* (*adaptive*) *cosmological model* does not seem preferable unless selection can preserve information that provides both a measure of continuity (a conservative tendency) and the capability to preserve those changes that promote cosmic stability (the adaptive or compensatory tendency) whenever they may emerge during *cosmic mutagenesis*. A universe whose specific parameters are consequent to a random roll of the cosmic dice at the onset of each cosmic cycle would be extremely unstable.

It is, nevertheless, a fundamental and inherent instability that drives the cyclic process, and each expansion phase represents a cosmic reaction that relieves this instability. Through adaptation, the universe becomes progressively more proficient at developing structure that promotes relative, albeit transitory, stability during its expansion phase. The role of black hole formation is in part to provide continuity in the cyclic process through the preservation of information. This eliminates the necessity of the following cycle to begin anew with random initial conditions and also ensures a measure of constraint on the course of *cosmic mutagenesis*. In other words, black holes are the *cosmogenetic* repositories of *heritable traits*. Moreover, these black holes lie dormant until complete cosmic collapse permits the re-emergence of these traits (subject to possible thermodynamically driven *mutagenetic change*) upon the ensuing cycle's re-expansion.

One goal of this paper is to hypothesize the means by which this *cosmogenetic* information is preserved and suggest a fundamental cause of the cosmic instability that drives the cyclic process. With this in mind, I postulate the existence of a previously undetected (perhaps fundamentally undetectable) short-range force functioning in opposition to the gravitational force with a range limited to or below the Planck length. In spite of the possibility that this *antigravitational force* (not to be confused with Einstein's cosmological constant) may be beyond experimental detection, this does not preclude the possibility that its existence may be inferred or mathematically supported. The inclusion of an *antigravitational force* may provide a necessary corrective factor fundamental to the formulation of a quantum theory of gravity. Among other things, it may provide a solution to the *singularity problems* confronting both general relativity and quantum mechanics. Its short range of operation may be the limiting factor in determining the minimum size a discrete unit of space may possess and may also explain why subatomic particles must have extension as opposed to being pointlike. By eliminating *singularities*, the existence of this force would have important consequences regarding black hole structure, as well as the structure of the universe at the moment of the *big bang*.

The existence of *dark stars* was first proposed in 1783 by John Michell and again by Pierre Laplace fifteen years later, but the discovery of the wave nature of light discouraged further interest in these *stellar curiosities* during the nineteenth century. Renewed interest in dark stars, now called black holes (the term for them coined by John Wheeler), did not reoccur until the twentieth century, with the publication of Einstein's theory of general relativity and the following work of Karl Schwarzchild. Einstein had demonstrated the particle nature of light with his earlier work on the photoelectric effect, and even though light *particles* were still regarded as massless, the extreme curvature of space (in accordance with general relativity) produced by black holes could prevent their escape from within the event horizon. Although he later received a Nobel Prize for his work on the photoelectric effect, the most significant paper published in Einstein's miracle year of 1905 was his paper on special relativity. Special relativity was important because it changed forever our notions about the relationship between space and time. It accomplished this by replacing separate and absolute conceptions of each, with a combined concept of absolute space-time. The concept of absolute simultaneity was demolished because it was demonstrated that the speed of light (in a vacuum) remains constant for all observers irrespective of their frame of reference. Consequently, because space and time are intimately connected and dynamic, any measure of motion through space or of the elapse of time will differ for two observers in constant relative motion with respect to one another.

The constancy of the speed of light for all observers guarantees that there are no absolute frames of reference that all inertial observers can agree on for the basis of their measurements. Each observer will have his or her individual set of space-time coordinates as a frame of reference, and each is equally justified in believing that he or she is at rest with respect to other observers (this may not be strictly true if Mach's principle is accounted for in a modified version of special relativity). Special relativity proclaims that motion through space and time are complementary. With any motion of a system through space, there will be a corresponding decrease in motion through time and vice versa. The primary reason that none of these bizarre effects (distance contraction and time dilation) on space and time were recognized before the advent of special relativity is because they do not become appreciable unless inertial systems approach relativistic velocity. One of the consequences of the imposition of a *cosmic speed limit* is that an exponentially increasing supply (approaching an infinite amount) of energy would be required for an observer to approach or attain it. Although many of the effects of special relativity have been confirmed experimentally, some of its conclusions are based on a classical conception of continuous space that require modification if space is in actuality *quantized in discrete units* (perhaps in units of volume of the Planck length cubed).

A *quantization of space in discrete units*, in which the limit may be governed at a subatomic level by the range of an *antigravitational force*, would place strict limits on the amount of contraction of space attainable in the direction of travel by an observer racing along at relativistic speed. Consequently, there would also seem to be a complementary limitation on time dilation if space is indeed discrete. These limitations on both space contraction and time dilation would suggest that there is a fundamental relationship between the measured value of the cosmic speed limit and the size of the *discrete units of space* (which may bear some relationship to the Planck length).

Furthermore, if space is discrete, there are also important consequences for limits on the curvature of space (for time dilation as well) that are attainable in accordance with general relativity. After all, general relativity is a theory that is built on the foundation of the classical concept of continuous space. Just as special relativity altered our concepts of space and time, general relativity changed the way contemporary science views gravity (an incomplete view as I will propose shortly).

The motivation for general relativity came as Einstein realized that Newton's theory of gravity was inconsistent with special relativity. For Newton, gravity is an attractive force acting instantaneously at a distance on any objects with mass. It acts with a strength inversely proportional to the distance between them and proportional to the product of their masses. This means that if the sun were to suddenly disappear due to a hypothetical, cataclysmic event, the earth would immediately deviate from its presently elliptical orbit. However, it takes light approximately 8 minutes to reach the earth from the sun traveling at an approximately 300,000 kilometers per second. From Einstein's point of view, the earth would not deviate from its current orbit until approximately 8 minutes after the sun's hypothetical demise, if the propagation of gravitational effects is restricted to a velocity consistent with the cosmic speed limit.

In addition, although Newton provided an explanation of how gravity works, he was unable to suggest an explanation of what gravity is. According to Einstein, gravity is not a force in a manner similar to other fundamental forces, but rather the gravitational field embodies the curvature of space-time in the presence of mass and energy. The presence of mass determines the curvature of space-time, and reciprocally, the curvature of the space-time geometry determines the path along which objects travel. Any body subject to the influence of a gravitational field moves along a trajectory that is the straightest line (a geodesic) allowed within the context of the relative curvature of the space-time geometry it encounters during its motion. Even the path of light is influenced by the curvature of space-time produced by large concentrations of mass.

One of the experimental confirmations of general relativity occurred in May 1919 when an expedition led by Arthur Eddington confirmed Einstein's prediction of the bending of light during a solar eclipse. General relativity was also demonstrated to be successful at better explaining the planet Mercury's anomalous orbit than Newton's theory. In addition, the observed red shifting of light spectra from celestial bodies provided a further, important confirmation of Einstein's theory. For the most part, general relativity has been an admirably successful theory, and yet, one cannot escape the idea

19

that it is incomplete because of its foundation on a classical notion of continuous space. This has resulted in tremendous difficulties in all attempts at reconciling general relativity with quantum mechanics. All attempts at reconciliation have resulted in the introduction of unwelcome infinities that reinforce the notion that apparently physicists are missing something. It would appear that in general, *nature abhors infinities*, and unless general relativity can be modified to eliminate its *singularities* (whether at the big bang or in black holes), a quantum theory of gravity will continue to remain elusive.

After all, one may characterize the singularities that are alleged to exist at the center of black holes and possess infinite density as having a *density in perpetual freefall* (which is what infinite density implies). This troubling aspect of singularities should make one wonder how it would be possible for the big bang to occur in the first place if this understanding of the concept of infinite density is correct. (Among other things, how can space-time expand its dimensions from a singularity unless there is a limiting distance at which an instability triggers expansion?) This suggests that some mechanism or force must be discovered that limits space-time curvature to prevent the existence of singularities. What is needed is a means by which a stable though dynamic foundational structure for space-time may be attained at the shortest distances consistent with quantum theory. A quantum theory of gravity should include Einstein's notion of space-time curvature, but it should also restore gravity's status as an actual force in the manner of the other fundamental forces.

During his quest to develop a new theory of gravity, Einstein experienced what he called his "happy thought," a thought experiment in which he imagined an individual falling from the roof of a house. It is this *happy thought* that provided the insight needed to arrive at the *equivalence principle* that became central to his theory of general relativity. In this moment of inspiration, it occurred to him that such an individual would not feel his own weight and therefore would have no sensation of acceleration in a gravitational field. It is important to note that the only indication this unfortunate individual may have that he is accelerating is the visual evidence that he appears to be descending at an increasing velocity, as well as the physical sensation produced by air resistance during his descent. Finally, our unwilling subject will have the unpleasant experience of encountering the resistance of the earth upon the termination of his descent.

It is resistance to gravitational acceleration that one may suggest is every bit as significant as acceleration itself, and recognition of this may lead to a solution of the *problem of gravitationally induced singularities*. All inertial bodies resist acceleration in direct proportion to their mass (*inertial and gravitational mass are equivalent*), and this is the reason that all objects regardless of their mass fall at the same rate in the presence of a gravitational field. However, the gravitational force is extremely weak and, under most circumstances, easily overpowered by the much stronger electromagnetic force (10 to the 42nd times more powerful). The electromagnetic force is instrumental in giving shape and structure to all baryonic matter. It is this force, functioning through the exchange of *virtual* photons that operate between the electrically charged particles, that constitute all baryonic matter.

The negatively charged electrons orbit around the positively charged nucleus and are restricted to particular orbits, in accordance with the Pauli exclusion principle. The electrical repulsion exerted between the electrons of separate atoms prevents these atoms from catastrophic collapse into one another, effectively resisting the much weaker gravitational attraction between them. Without this electrostatic repulsion, there would be no stable macroscopic structure, and our falling protagonist in Einstein's thought experiment might still be falling. Actually, in the absence of the resistance supplied by the electromagnetic force, man and earth together would probably collapse into an extremely compact object (this will be discussed in more detail later).

Unlike the electromagnetic force that is believed to have infinite range, the strong nuclear force is a short-range force. This force holds quarks together to form protons and neutrons (and other more exotic particles), and it is also responsible for *trapping* these nucleons within the confines of an atomic nucleus. In addition, because it is approximately 137 times stronger than the electromagnetic force, it is able to overcome the resistance of the electromagnetic repulsion acting between the protons *imprisoned* within the nucleus. The weak nuclear force is the other currently recognized short-range force and is important in reactions involving radioactive decay of protons and neutrons. During nucleosynthesis in stars, a proton will convert into a neutron with the release of a positron and neutrino, and the positron soon annihilates with an electron, producing gamma rays. The weak interaction is also involved in the radioactive decay of unstable atoms in which a neutron is converted into a proton and a simultaneous release of an electron and antineutrino (there is also alpha decay involving the release of alpha particles). The gamma radiation produced through the annihilation of electrons with the positrons produced by the activity of the weak interaction (within the sun) is eventually transformed into some of the sunlight that reaches earth.

All of the nongravitational forces have an important, composite role in the organization of the complex *hierarchical structuring* of matter and an equally important role in influencing the geometry of the *space-time fabric*. Without the activity of the aforementioned derived forces (which excludes gravity and antigravity), there would be no complex structure and no people to wonder about it. Gravity would have complete reign within the universe, and if all of the derived forces were to suddenly disappear, the universe would collapse in short order. There are places where gravitational collapse does occur when sufficient mass is concentrated enough to *overwhelm* the activity of the derived forces. It is already recognized that the *supermassive black holes* found at the center of galaxies have a fundamental role in early galaxy formation. If the *cyclic cosmological model* is correct, black holes probably have an even more significant role in cosmic evolution.

According to theory, if the mean density of matter and energy in the universe exceeds what is called the critical density, the universe will have positive curvature and is destined to collapse sometime in the future. However, if the universe has a previously unrecognized complex structure that has some bearing on how the collapse transpires, there is no reason the *big crunch* should happen all at once. In fact, if the space-time fabric has a *stratified hierarchical structure* of coexisting and weakly *interacting space-time geometries*, it is quite likely that cosmic collapse will occur in stages. If this is the case, then it may not be a question of whether the universe will collapse, but rather whether the *process has already begun*, albeit on a small scale. Every stellar and *supermassive black hole* may be circumstantial evidence supporting the notion that the collapse begins locally and ends with the final collapse of the vacuum at some time in the distant future.

It is precisely because complete gravitational collapse is contingent on the combined collapse of the weakly interacting, coexisting space-time geometries and their associated fields, that *collapse must occur locally and in stages*. Each of these hypothesized space-time geometries will have its own individual set of forces and associated parameters and is associated with the *three types of matter* that make up our universe. Baryonic matter, which constitutes approximately 4 percent of all matter, is to be associated with what will be called the baryonic stratum of space-time geometry (that is, baryonic matter oscillates into the baryonic stratum). In contrast, dark matter (approximately 23 percent of all matter) is composed of particles that oscillate into the dark matter stratum. Dark energy, which makes up the remainder (approximately 72 percent), is a product of the fields of the dark energy stratum and is the third of three strata that are *organized in a hierarchical relationship*. In addition to possessing an individualized set of parameters and forces, each layer or *space-time stratum* will have its own individual allocation of four space-time dimensions (three space dimensions and one time). Although all *three strata* may share some similar parameters and forces, there may be differences in parameter values and in the nature and strength of the forces operating within them.

These differences are reflected in the complexity of structure that is attainable within the constraints attributable to each stratum of space-time geometry. For example, dark energy strongly resists aggregation, and because it only weakly interacts with the other two types of matter, the dark energy stratum provides an ideal environment for the interaction of dark and baryonic matter to produce complex structure. Additionally, dark energy not only resists aggregation, but it also possesses a repulsive property that induces expansion of the cosmic environment. Although dark matter may have a combination of attractive and repulsive properties (because of the nature of the stratum-dependent forces that act on it), it is also somewhat resistant to aggregation and therefore unable to produce complex structure on its own. Dark matter only weakly interacts with baryonic matter gravitationally, but its presence is indispensable for the development of galaxy and galactic cluster formation (as well as stellar processes, as will be demonstrated later). The difference in properties and roles of the various types of matter is attributable to the *variation in multidimensional structure in which they oscillate.*

The notion of an *extradimensional structure* of the universe is not new. In a paper he sent to Einstein in 1919, Theodor Kaluza suggested the existence of an extra spatial dimension in an attempt to unify gravity and the electromagnetic force. A further development of this idea was proposed several years later by Oskar Klein, with the suggestion that this extra dimension would be curled up, circular, and so microscopically tiny that it would be virtually hidden from observation. The idea of extra dimensions lay dormant for decades until its resurrection in what became string theory and later M-theory. Both the string theory and the M-theory retained the notion of extra, curled-up, spatial dimensions, but as yet, no string theory has been developed that incorporates *extra time dimensions*. Ultimately, it is probably the failure of the string theory to accommodate *extra time dimensions* that has prevented it from achieving definitive success. Once this inclusion is accomplished, it may also be found that the extra dimensions are *not curled up*, but rather *out of sync* with those of the (four-dimensional) baryonic stratum owing to their *variance in parameters.*

The key to producing these *extra time dimensions* lies in a new application of Einstein's theory of special relativity that accounts for the *tri-layered hierarchical structuring of the space-time geometry*, as well as a stratum-dependent difference in measurement for the speed of light. (There is a *stratum and parameter-dependent variation in four-dimensional space-time.*) Stratum-specific values for the constant c will result in *time dilation variations* from stratum to stratum, which impact the motion of *strings* as they oscillate through the entire space-time geometry (the entire *multidimensional geometry* to which they are restricted that varies for each of the three types of matter). *Stratum-related variations in space contraction* will also have a significant relativistic influence on the motion of these *oscillating strings*. In addition, *the size of the smallest discrete units of space and time will be stratum-specific* as well, establishing a limitation on the contraction of space allowable for each stratum of space-time geometry.

As mentioned earlier, each stratum of space-time geometry has its individual allotment of fundamental forces with associated gauge fields. All of these *stratum-specific gauge fields* weakly interact with each other under the right circumstances, and quantum fluctuations may arise with constructive or destructive interference whenever they are induced to interact. These *interstratum gauge field interactions* occur primarily during high-energy reactions or whenever relativistic velocities are involved (as well as in the presence of strong gravitational fields). The gauge fields and related forces of the dark energy stratum are more *primitive* and more symmetric than those for dark matter. In like manner, the forces associated with dark matter will also exhibit greater symmetry and simplicity than those of baryonic matter.

Although the physics involving dark matter and energy is yet poorly understood, contemporary science has accumulated a great deal of understanding of the physics involved with baryonic matter. For instance, it is understood that stars form when a sufficient amount of hydrogen gas aggregates and condenses to the point that it collapses under gravitational pressure, triggering nuclear fusion. The star's gravitational pressure is balanced by the outward pressure produced by nuclear reactions in its core that causes it to achieve a state of hydrostatic equilibrium. In the star's core, protons from the condensed hydrogen gas are brought close enough together by gravity to overcome the electromagnetic repulsion between them and to become confined by the strong nuclear force. Helium is formed during this process, which involves the conversion of protons into neutrons and the release of energy (all of which will be discussed in greater detail later).

The projected lifetime of a star is contingent on the amount of mass it possesses, and larger or more massive stars tend to have relatively short lifetimes lasting only approximately 10 million years because they consume their fuel more rapidly than relatively smaller stars. The final fate of small (but not too small) and medium stars is to contract into *white dwarfs* at the termination of their life cycles. These white dwarfs are prevented from further contraction by the electrostatic repulsion existing between the stars' electrons due to electron degeneracy pressure as predicted by Pauli's exclusion principle.

The Pauli exclusion principle is also an important factor in the formation of neutron stars. In these stars, which are the end product of the evolution of some extremely massive stars, the gravitational strength of the core remnant is believed (I'll have more to say about that later) to cause the stars' protons and electrons to combine to form neutrons. The resulting neutron star is forestalled from further contraction by the neutron degeneracy pressure that places a limit on how closely the neutron star's neutrons can pack together. If, however, upon meeting its demise a massive star has sufficient mass after shedding its outer layers in a supernova explosion to overcome neutron degeneracy pressure (and quark degeneracy pressure), it will become a black hole. According to current theory, the core of this black hole remnant shrinks to an infinitely dense point called a *singularity*. The singularity is, in turn, surrounded by a boundary called the *event horizon*. Nothing, not even light, can escape once it has passed through this event horizon. Singularities are believed to have no measurable volume, possess infinite density, and occupy an infinitesimally small region of space that exhibits infinite curvature. *Black hole singularities*, as well as the so-called *big bang singularity*, are objects, the existence of which is predicted by general relativity in its current form.

However, if it can be demonstrated that general relativity is incomplete, it is not unreasonable to suppose that *singularities do not actually exist*. Moreover, this will have far reaching implications for cosmology and black hole physics. If space-time is *not continuous* and instead has a *discrete structure* (there is no space-time continuum if *space-time is quantized*), then such notions as infinite density and a clearly delineated first moment of time may make no sense. Instead, it is probable that space-time curvature and structure have established limits consistent with the *discrete quantization of space* (and time). These limits will also be consistent with the *level of hierarchical organization* and operating forces associated with each stratum of space-time geometry.

The relative expression or dominance of the *nongravitational forces* (derived forces*)*, as well as their role or function, appears to have a connection to their situational relationship to gravity. For example, while a star is in hydrostatic equilibrium, the relatively weak gravitational force is able to overcome the electromagnetic repulsion exerted on the protons contained in its hydrogen fuel. The energy produced in this process of stellar nucleosynthesis provides an outward pressure that counterbalances the effects of gravity. As long as the star has fuel to consume, it remains in a state of hydrostatic equilibrium. However, as a star ages, the role and function of the nongravitational forces changes as well, and this produces changes in the local space-time geometry. Upon their demise, stars attain *new states of equilibrium* that vary in accordance with their mass after their outer layers are shed.

White dwarfs exhibit an equilibrium state in which electron degeneracy pressure balances the effects of its gravity, and further contraction of the local space-time geometry is thwarted. Nevertheless, electron degeneracy pressure proves inadequate to the task of providing resistance to gravity in the case of neutron stars. These remnant stars are more massive and denser than white dwarfs and rely on the equilibrating influence of neutron degeneracy pressure (as mentioned earlier). Curiously, black holes appear to be the *odd man out* if current theory is correct. The *singularity hypothesis* would appear to suggest that there is no *equilibrium state of devolution* permissible to balance the effects of gravity. Nevertheless, if black holes do in fact exist in an equilibrium state, this cannot be accomplished through the action of the traditional nongravitational forces.

All of the equilibrium states just mentioned are attained through the operation of quantum processes, and it is certain that if a *state of black hole equilibrium* exists, it must be attained through quantum processes as well. The existence and operation of a *short-range oppositional force* would counterbalance the effects of gravity at or below the Planck length and *preclude the existence of singularities*. This *antigravitational* force may be fundamentally undetectable, but it may also have a fundamental role in the quantization of space by establishing a *barrier to infinite contraction*. The implication for cosmology is that if the big bang occurred, it could not have begun with an exponential expansion of space from a singularity. Moreover, if the *discreteness of space* is accompanied by a further *quantization of time*, then there may not have been a so-called first moment. If some version of the cyclic model is correct, this would indicate a possible measure of space-time continuity that establishes an unbroken connection between cosmic cycles.

If the universe never actually stops *to take a breather* (however, there may be a period of *localized dormancy* associated with black hole formation), then it must be due to some *inherent instability* that exists between the gravitational and antigravitational forces at or below the *Planck volume*. Ultimately, the bang (or the rebang) represents a disequilibrius reaction to the tremendous stress and pressure created by the unstable quantum *struggle* taking place between the opposing gravitational forces. The other so-called *fundamental forces* are merely the hereditary offspring of the oppositional gravitational forces and are the product of a long evolutionary history that spans innumerable cycles. This suggests that the gravitational and antigravitational forces are the only truly fundamental forces, and the derived forces (or offspring forces) are reduced to secondary fundamental status. By this, it is meant that the nongravitational forces remain fundamental within a restricted sense, which is tied to a specific cyclic and stratum context. On the other hand, the two primary fundamental forces retain their basic identity from cycle to cycle and never cease to operate, although their relative strengths and range may vary throughout the course of *cyclic evolution*.

In contrast, the secondary or offspring forces make their first appearance only during the *embryonic development* of a given cyclic expansion phase. None of these derived forces continues to operate inside black holes, although information concerning them is preserved by some means of *quantum imprinting*. Moreover, the reason light cannot escape a black hole is because the electromagnetic field is no longer active within the confines of an event horizon.

Both the gravitational and antigravitational forces permeate infinite space, and each discrete unit of space is subject to *stratum-dependent variation* in the strength and the range of both forces. The antigravitational force establishes a maximum limit on the space-time curvature (which is also stratum dependent) of every discrete unit of space. For the most part, the activity of this force is relatively insignificant, except in cases where sufficient mass is accumulated to promote quantum interaction with the gravitational force. Although the antigravitational force has limited range, it must have a *disequilibrius relationship in relative strength* with respect to gravity at their point of interaction. This may suggest that there is an *exponential increase in antigravitational strength* (producing a gravitational gradient) in opposition to the gravitational force at increasingly shorter distances. The measured value of the Planck length is dependent on the relationship between these two forces and is subject to variation, contingent on cyclic evolutionary history and stratum-specific differences in parameters. A stratum-related variation of parameters implies a commensurate disparity in stratum-specific measurements for the Planck length, as well as diversity in the range and the strength attributable to the *oppositional gravitational forces* operating in each stratum.

The attractive and repulsive properties manifested in the *parental oppositional forces* (the gravitational and antigravitational) are also characteristic properties that find combined hereditary expression in the offspring or derived fundamental forces. For example, in electromagnetism, a repulsive force is exerted between fermions of like electric charge and an attractive force is exerted between fermions of opposite charge. These repulsive and attractive characteristics are found, with varying degree of expression, in the nuclear forces as well. The electromagnetic force is believed to have an infinite range; however, this has not been proven, and there may be a cosmic structure-dependent limit that awaits discovery (this may also be true for the gravitational force). In contrast, the strong and weak nuclear forces have short ranges that are restricted to the nucleus of atoms.

Like the nuclear forces, the antigravitational force has a short range, and it is responsible for establishing the boundary of each discrete unit volume of space. At this *boundary*, the antigravitational force resists further contraction, and there is a discrete amount of this force present in each discrete space-time unit. There may also be a degree of elasticity present at each boundary, reflective of the antigravitational force's propensity for an exponential increase in strength (over and above the strength of the gravitational force) at increasingly shorter distances.

If the theory of supersymmetry is correct, it may be that the antigraviton (a boson with mass and hence short range) has a fermion superpartner that would be called the antigravitino. If so, it may be that an antigraviton is paired with an antigravitino, and these antigravitinos are confined within the volume of every discrete unit of space. It is possible that local space-time curvature could be determined by a distance-dependent relationship between the superpartner particles of the gravitational (gravitons and gravitinos) and antigravitational forces (antigravitons and antigravitinos). However, I have my doubts about the correctness of the supersymmetry theory and the proposed existence of superpartner particles. I believe that a complete understanding of the complexity of space-time structure may make such a theory unnecessary. In any case, the strength of gravitational attraction exerted between two bodies may be a consequence of its distance—and mass-dependent attraction to the antigravitational component of the local space-time geometry, or in other words, the *opposing gravitational forces* have an affinity for each other—an affinity like that of some lovers, growing the closer they become, as well as when there is increasingly more to *love* (wherever there is a concentration of mass).

The particles that compose an inertial body are held together gravitationally through the exchange of *virtual gravitons* (which will be discussed in further detail during the discussion on *virtual particles* later in this work). Although gravity may have an unlimited, or perhaps even a cosmic, structure-dependent range, virtual graviton exchange can occur over vast distances. However, the *short-range limitation* of the antigravitational force prevents the exchange of antigravitons (virtual or real), except at the periphery of each discrete volume of space. It may be that the relative weakness of gravity is due to a partial neutralization process that occurs at the periphery of each of these discrete units of space. In effect, a measure of gravitational strength is cancelled out through Planck boundary interaction with the antigravitational

33

force, which limits the shortest distance at which the gravitational force may operate.

As a consequence of its role in establishing the size of the Planck length and volume, the antigravitational force plays a significant role in determining the value of the gravitational constant, as well as the constant c. A corresponding decrease in the range of the short-range antigravitational force would produce a smaller value for the constant G (as well as for the reduced Planck constant), with a reciprocal increase in the value for the constant c. If the size of the Planck volume is stratum dependent, then there will be corresponding variations in the fundamental parameters that govern each cosmic stratum. This may also suggest further stratum-dependent variation in the nature of the derived forces that operate within each stratum of space-time geometry. These differences will be reflected in the nature and the complexity of structure permissible in each stratum and will also influence the manner by which *interstratum interaction* operates.

As a consequence of its influence on the size of the Planck volume, the antigravitational force provides a structural foundation on which the *extension* of particles is assured and a pointlike nature is invalidated. It also establishes a finite minimum limit to the range of the derived forces, as well as a barrier to the motion of subatomic particles at Planck or sub-Planck distances. For quarks, this suggests a minimum distance limit to their asymptotic freedom. This means that these subatomic particles have a range of motion restricted by both an outer barrier enforced by the strong nuclear force and an inner barrier enforced by the activity of the antigravitational force.

A successful inclusion of the antigravitational force in a modification of the standard model may be what is needed to *eliminate the singularities* that have forestalled successful attempts at the conciliation of quantum mechanics with general relativity. Moreover, a model that successfully incorporates this short-range force, as well as the hierarchical stratification of the space-time geometry, and all it entails may militate against the need for the renormalization and perturbation theory. It may also have profound consequences for theories of black hole structure and for the overall structure of the universe at the moment of the theorized big bang. Instead of collapsing to a singularity, it may be that the dead residuum of a collapsed star attains a state that is similar to that of *hydrostatic equilibrium.*

This equilibrium state would be a state of *gravistatic equilibrium* in which there is a cosmically transitory (for reasons to be dealt with later) balance between the oppositional gravitational forces. A prescribed limit on the amount of antigravitational force and information content attributable to each unit volume of discrete space implies that black holes may have a *quasicellular structure.* This suggests that a variation of the Beckenstein bound should be applied at this quasicellular level and that black holes possess volume that extends to the boundary of the event horizon. Information that becomes trapped in these quasicellular units is not lost but is preserved in a state of dormancy by some manner of *quantum imprinting.* This exhibits the *cosmogenetic* function of black holes in that they preserve information that may find re-expression and inform the development of subsequent cosmic cycles. Of course, this is all contingent on the correctness of the cyclic model. Current rejection of cyclic models by the mainstream scientific community is at least in part traceable to steadfast attachment to a rigid formulation of the second law of thermodynamics. Unfortunately, this ignores the possibility that the second law may have a limited contextual role in cosmic evolution (that it may have an intrastratum but not interstratum application). It also does not take into account a currently unrecognized *recycling capability* attributable to the mutual interaction of the oppositional gravitational forces at the quantum level.

According to current theory, the ultimate fate of the universe is dependent on the critical density of matter and energy present within it. This is represented by omega, which is significative of the ratio of measured density to the density that is a prerequisite for cosmic collapse. It is believed that if omega is greater than 1, then the universe is considered closed, and it will inevitably cease expanding at some point not yet determined and collapse. Nevertheless, if the value of omega is less than 1, the universe's projected fate is one of eternal expansion, and in this case, the universe is said to be open. The remaining alternative in which omega is equal to 1 pictures a flat universe—a universe that will expand eternally at a slower rate than if omega has a value of 1. Of course, these assumptions rest on what is now an incomplete knowledge of cosmic structure and, I believe, an incomplete knowledge of the totality of forces and their *interrelationship*, which may have a role in determining the ultimate outcome.

Experimental measurements of the cosmic microwave background radiation attained by WMAP (Wilkinson Microwave Anisotropy Probe) and BOOMERANG (Balloon Observations Of Millimetric Extragalactic Radiation and Geophysics) appear to support a model of a flat universe, at least on the surface. This would seem to be an apparent, at face value, support of the notion of an eternally expanding universe, expanding at a slower rate than if omega were equal to less than 1. Nevertheless, the actual density of the universe cannot be definitively ascertained at this time. This is in part because of the fact that dark energy and dark matter continue to resist experimental observation. Since the nature of both remains unknown at present, cosmologists are unable to specify the role each may have in determining cosmic fate. Furthermore, the existence of both types of matter has only been inferred through the manner in which each appears to influence cosmic structure.

The presence of dark matter has received apparent observational confirmation as a consequence of the work of Fritz Zwicky (1933) and Vera Rubin (1970), with additional support from gravitational lensing observations. Zwicky found that galaxies in the Coma Cluster were moving much faster than they should be, suggesting that most of the mass present in these clusters is unseen. Moreover, by studying the motions of stars within galaxies, Vera Rubin's group concluded that without the presence of a great deal of nonluminous matter (approximately 90 percent of a given galaxy's mass), many of the stars inhabiting a galaxy should not remain gravitationally bound because of their great speed. As a result of the evidence attained by these researchers and of that obtained through research on gravitational lensing, it is now believed that galaxies and galaxy clusters are enveloped within large *dark halos*. It is further believed that the dark matter comprising these dark halos has a consequential role in galaxy formation and structure.

The actual composition of this matter remains a mystery because of, in large part, the fact that it interacts weakly with what is conventionally regarded as normal matter. Even its interaction with light in gravitational lensing is not a consequence of direct electromagnetic interaction. Instead, the gravitational bending of light is believed to be due to the impact dark matter has on the local space-time geometry it occupies. In addition, the inability of dark matter to develop complex structure on its own is a key indicator that the parameters (nature's constants, etc.) that operate on it may differ significantly from those that govern the activity and motion of baryonic matter.

The history surrounding the concept of dark energy has an earlier inception than that of dark matter as a result of Einstein's reluctance to initially accept conclusions derived from general relativity, which he found unpalatable. This reluctance instigated the inclusion of what has become known as a *cosmological constant* into his theory on gravity to eliminate solutions that inevitably deviate from that of a *static* universe. As the story goes, this led to a later proclamation by the renowned scientist that this theoretical manipulation of his field equations was his *greatest blunder*. Curiously, the same Einstein who had postulated the dynamic nature of space-time required the observations of Edwin Hubble to provide convincing evidence of cosmic expansion to be disabused of the contemporaneously popular notion of a universe in stasis.

A reinvigorated incarnation of Einstein's rejected idea re-emerged from the *dustbin of history* after many years of ignominy, when observations began to accumulate, which seemed to indicate that the present expansion of the universe is accelerating. An expectation of a slowing down of cosmic expansion had, up until recently, achieved virtually universal concordance within the community of cosmologists. However, based on distant supernova measurements (type 1a supernovae are used as standard candles because of the fact that they seem to have identical intrinsic brightness), it has been determined that accelerated expansion of the cosmos began several billion years ago and has been increasing ever since. This unexpected *quirk* has led scientists to conclude that there is a *repulsive force* driving this expansion and that the vacuum is permeated with what is now called *dark energy*. To reiterate a point I made earlier in passing, dark energy is not to be equated with the antigravitational force, which has a significantly different role in influencing cosmic structure and development.

The antigravitational force, if it exists, should be regarded as one of the five fundamental forces (actually, I will suggest that there are stratum-dependent variations in this number) that should be separated into categories of *primary* and *secondary fundamental forces*. The oppositional gravitational forces comprise the primary category and function as inherent properties of space-time geometry, which may have stratum-dependent, as well as cycle-specific, variations in strength and range. These forces are also distinguished by possessing an unchangeable nature with respect to their individually attractive and repulsive properties. On the other hand, the *secondary category* is comprised of the derived forces that are the hereditary offspring of the primary fundamental forces. Unlike the primary forces, these secondary forces are not an eternally inherent property of space-time geometry and will vary in nature as adaptations to changes within a cycle-specific developmental or evolutionary context. This means that the repulsive and attractive properties of the parental forces will be transmitted in varying measure as *cosmogenetically inherited characteristics*. Moreover, these characteristics find expression in the offspring forces during post-big bang development as critical thermodynamic thresholds are attained. These critical thresholds will also be subject to cycle-specific variation in accordance with the adaptive changes however slight, which may alter the nature and properties of the derived forces.

In contrast to the short-range antigravitational force, dark energy is one of the three types of matter that inhabits the cosmos, with a function and operation that are primarily restricted to the bottom stratum of the hierarchically stratified space-time geometry. As is the case for the other strata, the particular measurement of the Planck length assigned to the dark energy stratum reflects the stratum-dependent variation in strength and range of the antigravitational force. There is also a related variation in the parameters that operate in the dark energy stratum, as well as a weaker gravitational attraction between its particles. As previously mentioned, the weakness of the gravitational force operating in any of the strata is the consequence of a neutralization of much of its strength at the shortest stratum-specific distances in interaction with the antigravitational force (henceforth, in most instances, we will call it the *Anti-G force* and the gravitational force will be referred to as the *G force*). Because much of the G force is used up in interaction with the Anti-G force at or below the Planck length (and in the creation of the offspring forces), less is available for exchange between the particles of matter that circulate within the three strata.

As was also mentioned, the measured value of the constant G has some bearing on the actual measurement of the Planck length. A shorter Planck length may suggest a smaller value for G and the reduced Planck constant, with a corresponding increase in measurement for the constant c. Therefore, a stratum-related variation in Planck length supplies a possible explanation for the weaker gravitational attraction acting between particles occupying the lower space-time strata. It essentially supplies a reason (although only partially in the case of dark energy) why the more *primitive species* of matter resist aggregation into complex organization.

Interestingly, a higher value for the constant c suggests an increase in the velocity at which this relatively weaker G force is transmitted between particles inhabiting these lower strata as well. Stratum-specific variation in key parameters effecting space-time geometry requires that there must be a stratum-specific application of both special and general relativity that accounts for these differences. There will also be some important consequences for quantum mechanics, which will be discussed later. A particularly significant consequence for Einstein's equation $E = mc^2$ is that because the constant c is larger for the two lower strata, E will have a correspondingly larger value for an equivalent amount of mass within these lower strata. In other words, an equivalent amount of mass of dark energy will have greater energy content than that of dark matter, and dark matter will have proportionally greater energy content per unit mass than baryonic matter.

One of the perplexing problems confronting modern cosmology is the apparent disproportionate ratio of normal matter to antimatter in the universe. Current models of the early universe suggest that there was a nearly equivalent amount of both matter and antimatter production during the big bang. It is believed that this ratio was not quite equal and that the remaining baryonic matter is a residue left over after the annihilation of most of the normal matter when it interacted with the primordial antimatter. However, I suggest that the majority of antimatter created in the early moments (although there was some annihilation during early cosmic development) has continued to remain with us all along. The reason that matter/antimatter annihilation is not observed on a regular basis, except under high-energy conditions is because the antimatter is normally confined within the dark matter stratum of space-time geometry. In other words, the large dark matter halos that envelop galaxies and galaxy clusters are in actuality large collections of weakly interacting antimatter. As has been suggested, this dark antimatter

41

operates within parameters that differ from that of *normal* matter. Therefore, it requires high-energy interaction between the baryonic and dark matter strata to produce the relativistic effects that permit brief *interstratum synchronicity* to materialize between them. In turn, this briefly imposed synchronicity between the strata produces equivalently brief interstratum symmetry to exist between the fields of force that govern each. Because of their mutual incompatibility, the symmetry existing between the strata is unstable and is quickly broken, resulting in the production of an antiparticle that is permitted to *live*, however temporary, in the higher stratum.

These released particles will exhibit a composite admixture of characteristics borrowed from each stratum but will be constrained by the particular forces prevailing in the higher stratum in which they now inhabit. For example, the positron released during nuclear reactions is produced as a result of symmetry breaking between the combined electroweak force operating within the dark matter stratum and the separate electromagnetic and weak forces operating in the baryonic stratum. This positron is composed of dark matter that has been altered through relativistic interstratum interaction in a manner that permits it to function in some ways like an electron.

Put another way, an energy contribution from the dark matter stratum is allowed to *exist* as an excitation of the electromagnetic field of the higher stratum (the baryonic matter stratum). This means that its activity will be constrained by some of the parameters that are associated with the higher stratum's electromagnetic field. This includes its higher stratum-dependent value for the constant c, which regulates the velocity of transmission of the electromagnetic force. In other words, photon exchange between two positrons released into the baryonic stratum will occur at the same velocity as it would between two electrons. However, the velocity of transmission of the combined electroweak force acting between any dark matter, which remains confined to the dark matter stratum, will exceed that of the electromagnetic force that operates within the baryonic stratum.

In the case of dark energy, its repulsive property is in part due to the comparatively weaker gravitational field prevailing in its stratum. It is also possible that the particles of dark energy possess a monopolar character that is currently repulsive. The existence of *monopoles* is predicted by theory, but their presence has not yet been detected by researchers. This is probably because the activity and the function of these *monopoles* has been restricted almost entirely to the dark energy stratum (mystery solved). The *broken symmetry* that is a consequence of stratum-dependent variation in parameters has also made detection more difficult. It may be that the monopolar character of dark energy is a property that is a product of a stratum-specific variation of the derived forces. In the dark energy stratum, s*ymmetry breaking* did not occur between the derived forces in the manner that it did in the other strata. The strong, weak, and electromagnetic forces persist in a combined form to make up the *strong-electroweak force*.

In addition to a variation of parameters, each stratum has its own family of fundamental forces associated with it. The number three appears to be favored by nature, at least in the current cosmic lifecycle. There are three families of fundamental particles associated with baryonic matter that are currently recognized. There are also believed to be three types of matter: (1) baryonic matter, (2) dark matter, and (3) dark energy. In keeping with this pattern, there may also be three strata of space-time geometry, each of which has its own family of fundamental forces. A proposed listing of these three families is presented in the following table:

(a) Dark energy stratum	(b) Dark matter stratum	(c) Baryonic matter stratum
Antigravitational force	Antigravitational force	Antigravitational force
Gravitational force	Gravitational force	Gravitational force
Strong-electroweak force	Strong force	Strong force
	Electroweak force	Electromagnetic force
		Weak force

Caption:

(a)The dark energy stratum has the smallest measured value for the Planck length and the weakest gravitational field, and the strong-electroweak force that has a short range and is weak compared with the derived forces of the other strata. It also has the largest value for the constant c. This stratum also has the shortest range antigravitational force, along with the smallest discrete unit of space. (b) The gravitational field is stronger for the dark matter stratum than it is for the dark energy stratum. This stratum also has a longer Planck length, and its variation of the strong force is weaker than that of baryonic matter. Because of this stratum's longer Planck length, there is a correspondingly longer range for its short-range antigravitational force. The combined (short-range) electroweak force is also weaker than the baryonic matter's electromagnetic force. In addition, the constant c has a smaller value than that of the dark energy stratum. (c)The baryonic matter stratum has associated with it the longest Planck length and the longest range for its short-range antigravitational force. Consequently, this stratum has the largest discrete unit of space associated with it. It also has the strongest gravitational field but the smallest value for the constant c. Each stratum has its own distinct value for Planck time and Planck temperature as well.

	4 dimensions	8 dimensions	12 dimensions
baryonic matter stratum			↑ particles oscillate thru 12 dimensions (9 space, 3 time)
dark matter stratum		↑ particles oscillate through 8 dimensions (6 space, 2 time)	
dark energy stratum	↑ particles oscillate thru 4 dimensions (3 space, 1 time)		
	dark energy particles	dark matter particles	baryonic matter particles

Dotted lines signify strata through which particles oscillate

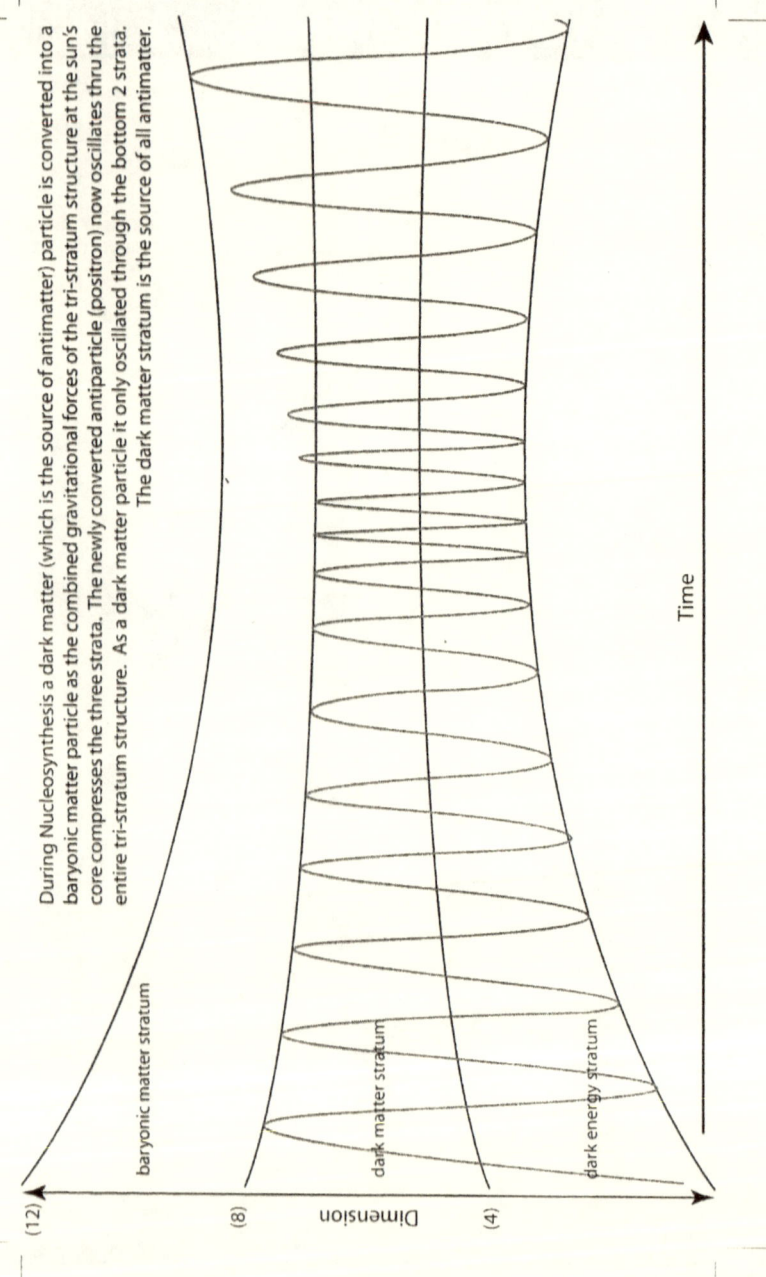

During Nucleosynthesis a dark matter (which is the source of antimatter) particle is converted into a baryonic matter particle as the combined gravitational forces of the tri-stratum structure at the sun's core compresses the three strata. The newly converted antiparticle (positron) now oscillates thru the entire tri-stratum structure. As a dark matter particle it only oscillated through the bottom 2 strata. The dark matter stratum is the source of all antimatter.

Time

Dimension

(12)

(8)

(4)

baryonic matter stratum

dark matter stratum

dark energy stratum

The variation in nature, range, and strength of the forces associated with these proposed families influences the structure of each stratum's space-time geometry. As the preceding table indicates, each stratum has assigned to it its own discrete unit of space and a corresponding limit on the range of operation of its antigravitational force. String theories, in general, postulate that there are ten space-time dimensions and that nine of these are space dimensions, and yet there is only one dimension of time. The M-theory adds one more space dimension to its total, but this is not important for the current discussion.

One of the difficulties connected with string theories is their inability to provide a convincing explanation why three of the space dimensions expanded at the expense of the remaining ones. This problem may be avoided if there is a hierarchical grouping of dimensions in sets of four (three space dimensions and 1 time), with each group attached to a specific stratum of space-time geometry. Since each stratum has its own Planck length and associated measurement of absolute space-time, this establishes the relationship that exists between the dimensions of the strata that constitute the space-time fabric. The lower groupings of space-time dimensions belonging to the dark matter and dark energy strata are not curled up. Instead, the stratum-dependent variations in Planck length, parameters, and forces operating in each stratum limit the interaction between the four-dimensional groupings of space-time geometry. (The introduction of three stratum-dependent variations in Planck time permits the inclusion of extra time dimensions in the extradimensional space-time geometry). This also impacts the motion of strings (if they exist) as they vibrate through the entire tristratum structure, and this will be discussed in more detail later.

There are important implications for black hole formation as a consequence of the variation in size of Planck volume associated with each stratum. Black holes are the end product of gravitational collapse in a region of space. If this collapse includes localized collapse of the entire tristratum structure, then the quasicellular units of space-time that black holes consist of, will come in three distinct sizes. In addition, if there is a relationship between the amount of information that can be contained in a discrete volume of space, then it may suggest a related variation in the complexity of information content as well. This means that the information preserved by quantum imprinting may contain a record of the forces (and their fields) that operated within each stratum in the effected local region before collapse. When a black hole forms, there is a rapid cascading collapse of the three strata beginning with the baryonic stratum and ending with the dark energy stratum. This also involves a cascading collapse of the derived fields associated with each stratum that renders them inactive until the next cosmic lifecycle.

Black holes are hierarchically organized in a state of gravistatic equilibrium, and they exist as a sort of *pinch* in the space-time fabric wherever they are located. As repositories of *encoded information*, they are transcyclic transmitters of cosmogenetic instructions that inform the development of an ensuing *cosmic lifecycle*. The three space-time strata are homologous structures, and the division of space-time into these variants of cosmic structure is probably due to some *mitotic-like* (or fissionlike) *process*. Moreover, it is likely that this *cosmomitotic process* requires a significantly longer span of evolutionary history than that required for the relatively simpler *parametric mutagenesis* of stratum-specific parameters. Nature appears to rely on extensive repetition of pattern in evolutionary processes. Therefore, it should not be a surprise that *homologous structure* is among those patterns that has a more universal application.

Our present universe exhibits an *interdependence* and measure of stability, which represents an adaptation to prior cyclic instabilities. As an adaptation, the present diversification of structure and function is the product of a multitude of incremental changes, each of which arose in response to contemporaneous instabilities. These incremental changes need not accrue during the transition from one cosmic cycle to another. Instead, it is likely that in a manner similar to that of biological evolution, consecutive *generations* (or in this case, cyclic lifecycles) will bear a striking resemblance to one another. Changes, when they do occur, probably arise when critical thermodynamic thresholds (reference: the Bénard cell experiment) are reached after a gradual transcyclic accumulation of thermodynamic stress.

In the course of what may be an eternal process of cyclic evolution, it is possible that there exists a *simplest* incarnation or form that the universe may assume. This does not suggest that there is a *cycle one*, in the strictest sense, because the cyclic process may be cyclic as well. Cosmic history may encompass an endless repetition of *cycles of identical cycles*, each of which begins with an identical initial cycle. Moreover, these initial cycles may follow in succession the final cycle of the previous grouping of cycles, which had itself attained a maximum threshold of complexity. By this, it is suggested that there is a limit to the degree of attainable complexity (for the architecture of space-time structure), which is dependent on the fundamental nature of and relationship between the oppositional gravitational forces.

Once this maximum limit is reached, the cosmos does not have the option of reverting to the lower level of complexity reached in a preceding cycle from which it minimally differed. After all, the final cycle in the chain of succession represents the evolutionary limit of complexity (which may represent a thermodynamic limit) of space-time structure. The cyclic variant that had preceded it became *obsolete* (selected against) because it had become thermodynamically unstable and ceased to be efficacious in relieving the instability between the oppositional G forces (there is a transcyclic thermodynamic connection between succeeding cycles). Instead, in keeping with its conservative evolutionary tendency, the universe would revert to the simplest conceivable form, which is permitted by the thermodynamic relationship that exists between the oppositional gravitational forces. (An analogy, perhaps a poor one, may illustrate what is meant here. Thermodynamic conditions in an environment may change sufficiently to cause the extinction of higher life forms but permit the existence of the most primitive microorganisms—a situation that may have occurred on Mars.)

A skeptic of the previous argument might suggest that the final cycle (that is the one with maximal structural complexity) alluded to might evolve in an infinite number of ways, which would be consistent with its specific level of complexity. However, nature appears to favor quantization over infinite variety, as the lesson provided by the solution to the *ultraviolet catastrophe* demonstrates. This means that simple variation of a similar nature may not be sufficient to relieve the thermodynamic pressures that render the maximum level of complexity unstable. At least this is true if these pressures are being brought to bear at the fundamental structural level (quantum level), upon which the architecture of this optimum manifestation of complexity is built, or put another way (borrowing a metaphor), cosmetic changes to the roof and walls will not compensate for structural stresses within the foundation of a house. The house will inevitably collapse like *a house of cards*, which will require that the builder begin anew with a new appreciation for structural limitations. Of course, the analogy is not perfect because the universe has no builder nor does it possess the sentience that would permit it to adopt a more moderate *plan* of construction that falls somewhere between both extremes of relative complexity. No, our universe is a *dumb brute*, and its only recourse is to revert to a cosmic plan that exhibits the least *imagination* and requires *minimal effort*, consistent with its energy resources and energy recycling capabilities.

Therefore, if there is a most primitive *cyclic incarnation*, designated *cycle one*, initiating a specific grouping, it would possess the simplest conceivable structure. As a consequence, it would most likely be constituted with a single stratum of space-time geometry and a single grouping of four dimensions (three space dimensions and one time). This simple stratum would be the ancient forbearer of our present dark energy stratum. However, unlike our present dark energy stratum, this early incarnation possessed an incredibly strong gravitational field, and a correspondingly powerful antigravitational field. At this initial stage of evolutionary development, there did not exist any derived forces to provide structural impediment to the interaction of the oppositional gravitational forces. The constant c did not represent the speed of light at this time, but rather the velocity of propagation of the two primordial forces, and it can be assumed that this velocity was incomprehensibly high (although not infinite) compared with the value of c that we are familiar with. Of course, this incredibly large value for the constant c would mean that the primordial Planck length and associated Planck volume would have been significantly smaller than present values. A significantly smaller Planck volume means a more rapid exponential increase in antigravitational force (Anti-G force) at increasingly shorter distances, far above that of any stratum-specific manifestation of this force operating at present.

The *thermodynamic gradient* that existed at this time would have been incredibly steep, and its reduction could only be achieved through rapid expansion. The rapidity of the expansion was a consequence of the tremendous variation in strength and range attributable to the primordial forces at increasing distances, as well as the incredible velocity of transmission of both forces. *Gradient reduction* at this early stage of cyclic evolution was quite inefficient at relieving the instability which characterized the relationship between the *Anti-G* and *G forces*. Although rapid expansion may have been effective at providing nearly instantaneous *relief*, this condition would have been short lived. This is because the rapid expansion would have been followed by an equivalently rapid collapse and a return to unstable conditions.

At this time, the single-stratum space-time fabric would have been oscillating back and forth, perhaps through a multiple number of generations with no significant or very little change, until a thermodynamic threshold of instability had been attained. The first changes or adaptations which emerged probably involved variations in the strengths, ranges, and velocity of propagation of the Anti-G and G forces. These changes in parameters would accompany changes in the size of the discrete volume of space and alter the relationship between the primordial forces. This would, in turn, impact the maximum temperature (it would be lower than it had been previously) attained during the initial conditions presiding in the first cosmic generation in which they emerge.

As a consequence of these changes, a discrepancy in the rates of cosmic expansion and collapse would begin to accrue. This time disparity is exhibited as an incremental increase in the time it takes for the initiation of collapse to commence in ensuing cycles. This will be in conjunction with corresponding variations in parameters and discrete volume size. With the incremental attainment of new thermodynamic thresholds over the span of many cosmic lifecycles, the universe begins to exhibit greater complexity of development and efficiency in *gradient reduction*. The *mutagenesis* of parameters that produces this greater complexity and efficiency is not Lamarckian in nature, but rather the consequence of spontaneous change in response to thermodynamic pressures. Eventually, in response to these pressures, it became necessary for the universe to modify the relationship that existed between the Anti-G and G forces in a more significant manner. This was accomplished through a spontaneous reproductive process from which emerged a primitive variant of what we now call a derived force. With its birth emerged a new set of spontaneously generated parameters that altered the space-time structure of the single-stratum universe. This new offspring force retained within it the characteristic attractive and repulsive properties that it had received from its *parental oppositional forces*.

With the emergence of this new force, the universe increased its ability to produce complex structure, and it became even more efficient at gradient reduction than it had through the process of *parametric mutagenesis* of the oppositional G forces. Over the span of many cosmic lifecycles, parameter modification and the operation of the single derived force proved to be an effective means by which the single stratum universe was able to attain increasingly longer periods of relative stability. However, the efficacy of both *parametric mutagenesis* and the functional operation of this new force eventually ran its course. A single space-time stratum could no longer suffice to produce cosmic stability. Consequently, in response to thermodynamic pressures, the universe adopted a new evolutionary tactic to relieve cosmic instability.

This was accomplished through a mitotic-like (or fissionlike) process (borrowing terminology from biology once again), which resulted in the division of space-time geometry into two separate strata, each of which retained its own variant form of the Anti-G and G forces operating within it. Each of these *homologous strata* also retained within it its own variant of the derived force, with differences that reflected the role-dependent nature and function of each force. Within the more primitive dark energy stratum (an early progenitor of our own), the derived force exhibited *monopolar characteristics* with a capability to reverse monopolarity at a late stage of a lifecycle's evolution. This end-of-life reversal reflects a change in relationship that the dark energy stratum's derived force has with the oppositional G forces as the universe ages. What occurs is a switch from a repulsive to an attractive tendency that alternately reinforces, first, the Anti-G force, promoting cosmic expansion, and then the G force as the universe approaches its expansion limit.

In the newly evolved dark matter stratum, the derived force adopted a role in which a comingling of the repulsive and attractive traits inherited from the parental forces produced an early version of a weakly interacting bipolar field. Although the particles produced as excitations in this field were themselves monopolar, unlike the lower-stratum form, these exhibited a rapid oscillation of repulsive and attractive properties. None of these oscillating monopolar particles possessed an individually identifiable electric charge. This is because each would have exhibited rapidly alternating negative and positive electric charge properties. These would have been unlike electrons and protons, which have identifiable negative and positive electric charges, respectively. For protons and electrons, the repulsive and attractive properties are interchangeable manifestations of the electromagnetic field, dependent on whether the nature of the interaction is between like or unlike particles.

The spontaneous appearance of the dark stratum during this early lifecycle would have acted as a brake, initially slowing down expansion of the early form of the dark energy stratum. This would have had an effect on cosmic lifespan, increasing it, and this would indicate that the *cosmomitotic division* into separate weakly interacting strata bestowed on this *cyclic variant* a selective advantage. An increased lifespan would have prolonged the time span between expansion and collapse phases, as well as the unstable interaction of the oppositional G forces.

The dark energy stratum assumed a new role in this new cosmic scheme as an environment within which the dark stratum could form loosely aggregated organizations of primordial dark matter. These aggregations would have formed via the clumping of the oscillating variant of monopolar dark matter particles. Moreover, this clumping would have become increasingly localized as the dark energy stratum continued to expand. Furthermore, this increasing localization of the concentration of the primordial dark matter would have limited the gravitational influence of this portion on the dark energy stratum. At some point during this cosmic lifecycle, the expansion of the dark energy stratum began to accelerate at an ever-increasing rate. This, in turn, increased the isolation of the widely dispersed aggregations of the primordial dark matter.

Although the dark stratum's derived force permitted only weak interaction between the locally aggregated particles, sufficient accumulation allowed the relatively weaker dark stratum variant of *G force* to assume local dominance. This resulted in a cascading collapse of, first, the dark matter stratum, followed by the collapse of the dark energy stratum trapped within these local regions of space. This was brought about as gravitational reinforcement between the strata, both overpowered the repulsive property and reinforced the attractive property of the dark matter's derived force. Gravitational collapse of the higher stratum then induced collapse of the dark energy stratum as the monopolarity of its derived field was forcibly reversed from its repulsive to its attractive property. The limited range of the Anti-G force prevents it from direct involvement in this collapse process, and its presence is only felt at its boundary of operation within each stratum.

The only role the Anti-G force has within the structure of these early black holes, as well those of later cosmic generations, is to assist in producing *gravistatic equilibrium*. This state of gravistatic equilibrium is brought about through the intercession of the structure of the collapsed, derived field (and later on fields that act as a sort of buffer in their collapsed state) as it prevents unstable interaction between the oppositional gravitational forces. The result is a relatively temporary (on a cosmic scale) local balance between the forces. Moreover, this balance will not be disturbed until the remainder of the dark energy stratum ceases production of dark energy (which will be explained later), inducing its complete collapse. With this final stage of collapse, one cosmic cycle ends and the new one begins. A cyclic generation's demise follows the reversal of the monopolar property from repulsive to attractive within the dark energy stratum (after the aforementioned cessation of dark energy production) via interstratum interaction of the G force. It is noteworthy that this reversal mimics the reversal of *monopolarity* brought about via the G force when it induces collapse of the dark energy stratum during black hole formation.

The mitotic-like division into two separate but interdependent strata represented a successful adaptation. Whatever evolutionary changes that accrued during the span of ensuing cosmic generations were intrastratum in nature. These included incremental mutagenetic changes in parameters, as well as gradual modification of the derived fields within each stratum. There were also corresponding intrastratum modifications in the range and the strength of the oppositional G forces, as well as a related intrastratum modification in discrete volume size. Changes in the dark matter stratum were more complex, involving greater mutagenetic modification of its parameters. There was also an incremental modification of the combined derived force inherited from the dark energy stratum during cosmomitotic division. This led to a mitotic-like division of the derived force's field into a progenitor representative of the strong force and an early form of what became that of the electroweak force.

This produced a *division of labor* that represented a diversification of function that enhanced the aggregation potential of the early form of dark matter particles. It was during this diversification process, which spanned the course of many cosmic generations, that separation into categories of particles that exhibited either positive or negative electric charge took place. This modification promoted clumping, which permitted localization of smaller accumulations of dark matter to accrue. This permitted greater separation of smaller concentrations of dark matter as the dark energy stratum expanded. Before the advent of the modification that introduced electric charge, dark matter accumulated in large clouds that had increasingly less influence on inhibiting the acceleration of expansion of the dark energy stratum. The wide dispersal of smaller accumulations of dark matter provided the means to decelerate expansion and prolong the lifespan of this cosmic generation. It also had the effect of slowing down the process of black hole formation, which allowed for the development of a more complex structure.

The early variant form of the electroweak force was relatively weak compared with that of later cosmic generations. However, it was quite effective at slowing down the process of black hole formation until there was sufficient accumulation of mass to overcome it. The Coulomb barrier, as we know it, did not exist until after the dark matter stratum underwent a further mitotic-like division, producing the baryonic stratum of space-time geometry. In addition, this Coulomb barrier did not arise until after the electromagnetic component of the electroweak field had evolved sufficient strength after mitotic separation. This was only permitted when there were sufficient incremental mutagenetic changes in the parameters operating within both the dark matter and the baryonic strata. Also required was an accompanying interdependent evolution in the variation of the strength and the range of the oppositional G forces operating within both strata as well. Any change in nature, range, or strength in a given stratum's derived fields is closely connected to the evolutionary changes affecting the relationship between these oppositional G forces.

Moreover, all changes, regardless of whether they are minor mutagenetic modifications of intrastratum parameters or the more consequential cosmomitotic division of one stratum into two, are thermodynamically driven processes. Each evolutionary change will impact the temperature at which the breaking of symmetry produces the spontaneous appearance of a stratum's derived fields and their further differentiation. It will also impact the temperatures at which the cosmomitotic division into separate strata occurs during early cosmic development. Adaptational changes to the relationship between each stratum's oppositional G forces affect the manner in which energy is recycled and distributed during this symmetry breaking process. Gravitational collapse of the remainder of the dark energy stratum at the end of the cosmic lifecycle induces interaction between the Anti-G and G forces. This induced interaction is spread throughout the entire dark energy stratum, including that part which had been trapped during black hole formation during the development of the cosmic cycle.

It is during the gravitationally induced interaction between the oppositional forces, throughout the collapsed dark energy stratum that energy recycling first takes place. The entropy produced in the previous generation had already been erased via interstratum processes before cosmic collapse (which will be explained later) and is transformed into useful energy through a recycling process. It is this newly transformed energy that reanimates the collapsed derived fields contained within each stratum. This *reanimation* occurs following the inducement of unstable interaction of each stratum's oppositional G forces that disturbs the pre-existing *gravistatic equilibrium* of the tristratum structure trapped within black holes. Each stratum trapped in these black holes initially suffers further collapse and then rebounds as the induced interaction between the Anti-G and G forces infuses recycled energy into each of them. There follows a cascade of reanimation of each of the three stratum's derived fields as each stratum is infused with recycled energy. Over the span of a number of cycles, variation in energy distribution during the final stages of cosmic lifecycles impacts the energy recycling process that initiates cosmic re-expansion. This leads to changes in the temperatures at which each stratum emerges during early cosmic development in ensuing cycles. Parameter values and the nature, range, and strength of the derived fields in the strata are also impacted by these incremental thermodynamic changes. This may lead to slight changes in parameter values or *parametric mutagenesis* and a spontaneous mitotic-like division of a stratum's derived fields. More significant thermodynamic changes occurring over an even greater span of cosmic cycles, eventually leading to a more complex form of cosmomitotic division that results in an incremental increase in the number of strata of space-time geometries, constitute the cosmic architecture.

With each mitotic-like division, there are also corresponding changes to the nature of the matter that are produced by the derived fields operating within the newly separated and homologous strata. The previous division of the dark energy stratum gave birth to the dark matter stratum within which the property of distinct electric charge was introduced. With the later cosmomitotic division of the dark matter stratum came the innovative polarization of structure, forming matter into separate stratum-specific types that are by their nature incompatible. Post mitosis, the dark matter stratum became the exclusive producer of antimatter and the newly formed baryonic stratum became the exclusive producer of what is commonly referred to as normal matter.

Under normal circumstances, these two incompatible types of matter only weakly interact gravitationally because of stratum-specific variations in parameters and derived fields.

It is only during relativistically induced interaction between the two strata that the dark stratum's antimatter is permitted to accomplish *trans-strata emergence* into the baryonic stratum. Once *trans-strata transduction* has been attained, the newly formed antimatter particles will function in a manner consistent with the derived fields and parameters present in the baryonic stratum. This means that they will behave in a manner identical to that of their fermion counterparts, which are native to the baryonic stratum. However, because these particles retain their incompatibility as a consequence of their opposite electrical charges, their existence within the baryonic stratum is normally short-lived. Whenever these antiparticles interact with their fermion counterparts, both are subjected to mutual annihilation, which results in the production of high-energy gamma rays.

The cosmomitotic division of the dark matter stratum, which gave birth to the baryonic stratum, proved to be a successful adaptation. It provided a selective advantage because it bestowed on the cosmos greater *thermodynamic efficiency* in accordance with the *law of maximum entropy production* (LMEP). There appears to be two oppositional tendencies operating in evolutionary processes. On the one hand, there appears to be what one might call a *compensatory* (sometimes overcompensatory) *tendency*, which is exhibited with the introduction of adaptations and complexification of structure. On the other hand, there also appears to be a strong *conservative tendency*, which is exhibited by the retention and incorporation of primitive structure within the formation of more complex organization.

The human brain is an excellent example of the tandem operation of these two tendencies. It consists of three coexisting and progressively complex organic structures: (1) the reptilian brain, (2) the premammalian, and (3) the neomammalian brain (Koestler and MacLean), which form a more complex and interdependent whole. The reptilian brain is the most primitive component of this organic whole and closely resembles that of the entire brain that functions within contemporary reptiles. In this sense, it has survived relatively unchanged because it still provides a vital function in the overall complex organization of the human animal (in the words of Koestler, "it has . . . become a stable evolutionary holon"). In like manner, the relatively primitive dark energy stratum has been retained with relatively little modification through the span of numerous cosmic cycles. It only went through a significant adaptation of role and function (a primarily environmental one) with the mitotic-like emergence of the dark matter stratum. In a manner analogically similar to that of the human brain, the tristratum structure of the universe forms an interdependent hierarchically organized architecture.

Just as the neomammalian brain represents the most recent addition to human brain structure, the baryonic stratum of space-time geometry represents the most recent addition to cosmic structure. The relative stability of this higher stratum is dependent, to some extent, on the relatively static evolutionary development of the more primitive dark energy stratum that functions as its environment. Within an ecosystem, extreme changes in an environment can negatively impact the survivability of a given species. It is also true that pathological changes to the human equivalent of the reptilian brain can disrupt the normal functioning of the higher neomammalian brain. In a similar manner, dramatic changes in parameters operating within the dark energy stratum can disrupt the stable interdependent interaction of the hierarchically integrated cosmic organization. Consequently, selection will favor the relatively static transcyclic persistence of the dark energy stratum in its primitive form with minimal modification spanning numerous cosmic generations.

Only upon the attainment of a critical threshold of trans-stratum instability may the dark energy stratum develop significant incremental change in its stratum-specific parameters. These spontaneous changes will be associated with corresponding changes in the relationship between the Anti-G and G forces and intrastratum fundamental (derived) forces. In addition, mutagenetic modification of its parameters (via parametric mutagenesis) may induce a cascade of mutagenetic change of parameters and discrete volume size within the other strata as well. The mutagenetic modifications of upper strata parameters will be induced because the lower stratum's transformation will impact the temperatures at which spontaneous emergence of the upper strata takes place. If thermodynamic pressures are sufficient, there may also be a further mitotic-like division of the contemporaneously existing superstratum.

Accompanying the spontaneous emergence of the new stratum will be an *interdependent* redistribution of role and function between this stratum and the substratum from which it was cosmomitotically produced. Eventually, the baryonic stratum will undergo mitotic-like division after the span of a number of cyclic generations following the present one, when thermodynamic pressures induce it. If the cyclic cyclic model is correct, that is, if there is cyclic repetition of identical groupings of evolving cyclic generations, there will be a limit on the number of strata that can be cosmomitotically generated. Ultimately, this will depend on whether there are minimum and maximum limits on possible Planck volume size. This, in turn, will depend on whether or not there are absolute evolutionary limits on the minimum and maximum distance ranges at which the Anti-G force may operate. The unstable relationship between the Anti-G and G forces produces the thermodynamic pressures that drive the cyclic process. The inability of these two forces to attain a permanent state of equilibrium between them is the ultimate source of impermanence and, at the same time, the source of novel adaptation of structure.

In fact, *Homo sapiens* owe their existence to this inherent defect, which plagues the cosmos at the shortest distances. Moreover, if there is a repetition in cyclic history from one cyclic grouping of identical cycles to the next, our species will make an infinite number of reappearances. That means that I will be sitting before this keyboard laboriously typing these words countless times. Of course, because of probability, slight variations may accrue from one cyclic grouping to another. However, probability does not rule out that an eternally existing, universe may eventually repeat the pattern of cyclic evolution, which produced those of us alive today. This may be the next best thing to an *afterlife*, although this may not provide much comfort to those who desire entry into a mythical paradise and the reward of eternal life.

The irreconcilable conflict between the Anti-G and G forces, which are fundamental to space-time geometry, ensures that the universe will be eternally evolving. The discreteness of space and the ubiquitous presence of the oppositional G forces may also suggest that the *universe has infinite extension*. Furthermore, it also rules out the existence of singularities, which means that the cosmic expansion could not have initiated from a single point. If the universe is finite and bounded, then there must be a finite amount of discrete units of space-time from which the big bang was launched. This might appear to suggest that the universe began as a tightly compressed *ball* of discrete units of space-time. Instead, it would seem preferable to hypothesize that the universe is composed of an infinite amount of discrete units of space-time and that the big bang represented the expansion of infinite space from a maximally contracted state. Expansion or contraction of an infinitely extended object only impacts its density but not its infinite character (space always remains infinite; in extent, there is only changes in the expansion and contraction of its multidimensional structure). For example, an infinitely extended number line remains infinite regardless of how it is divided or whether it is expanded or contracted. The same is true if we substitute our number line with an infinitely extended three-dimensional graph.

The difference between these examples and actual space-time geometry is that each of the infinite-in-number discrete units of space represents a point of origin from which finite expansion is initiated. If these discrete units vary in size, as determined by the boundary established by the strength and the range of the Anti-G force, there will be a corresponding stratum-specific variation in the temperature at which expansion begins. This will impact the relative timing (affecting changes in stratum-dependent variations of Planck time) of temperature-related spontaneous regeneration of the cosmic strata. Because the parameter c has the largest numerical value within the dark energy stratum, the Planck volume associated with it will be the smallest. For the same reason, the Planck temperature associated with this stratum will also be the highest. Accordingly, the baryonic stratum will have the largest Planck volume associated with it, as well as the lowest Planck temperature.

The unfolding of initial cosmic expansion begins as the Anti-G and G forces within the dark energy stratum are induced into unstable interaction upon the final collapse of this stratum at the end of the preceding cycle. The portion of the dark energy stratum that had been *entrapped* during the preceding cycle's black hole formation is also subjected to this unstable interaction. As mentioned earlier, this, in turn, disrupts the *gravistatic equilibrium* of the upper strata (or superstrata), inducing a cascade of unstable interaction of the Anti-G and G forces present within each.

During the course of this induced tristratum interaction, the Planck temperature of the dark energy stratum is rapidly obtained, and re-expansion is initiated. This is followed by spontaneous re-expansion of the dark matter stratum as its relatively cooler Planck temperature is obtained. The baryonic stratum has the *coldest* Planck temperature (still an unimaginably hot 1.41679×10^{32} K) associated with it, and consequently, it is the last stratum to spontaneously re-emerge. The rapidity of expansion initiation of each stratum is a consequence of each stratum's value for the constant c. Both the Anti-G and G forces are transmitted at this velocity, and this impacts the speed at which critical instability is attained as the result of induced interaction between these two forces within each stratum. In addition, the stratum-specific velocity of the constant c probably represents the velocity of initial expansion of each stratum upon the attainment of critical instability. This would suggest that the attainment of critical temperature was the most rapid for the dark energy stratum. It would also suggest that this stratum was able to expand to great size before the re-emergence of the upper strata.

As was mentioned earlier, black hole formation during the expansion phase of the cosmic lifecycle involves the collapse of all three strata in local regions of space. The key point to remember is that only a relatively small portion of the dark energy stratum is collapsed during this phase of cosmic collapse. The entire collapse of the dark and baryonic matter strata has been accomplished in localized regions of space with the termination of cosmicwide black hole formation. This is followed by the cessation of dark energy production, which leads to the final collapse of the remainder of the dark energy stratum. The quantization of space, in the form of stratum-specific discrete units, permits separation to exist between those regions of space, which contain the collapsed remnants of all three strata.

This means there will be no uniformity in the collapsed cosmic structure before re-expansion. Localized regions of the collapsed tri-stratum structure will already be dispersed before the initial expansion of the dark energy stratum. With the reanimation and the relativistic expansion of this stratum, the previous miniscule separation between the collapsed regions of the tristratum structure will be dramatically increased. In addition, the ensuing orderly reanimation of the upper strata will introduce a braking effect, as these re-emerged strata begin to gravitationally influence the lower substratum. This will occur after gravistatic equilibrium has been destabilized within the *upper strata*, permitting both to gravitationally interact with the dark energy stratum. The lower substratum will proceed to expand at a much slower and steady rate as the other two strata begin to construct complex structure in localized regions of space. As the universe continues to develop, these regions exercise, slowly decreasing gravitational influence on one another. Eventually, after the passage of billions of years, a critical point will be reached and expansion of the dark energy stratum will accelerate. Dilution of the influence of the upper strata brought about by increasing separation is only one of the reasons for this acceleration. Another reason for this acceleration will be explained later in this work.

In any case, it is possible that the aforementioned scenario takes place throughout infinite space within a single infinitely extended universe. It has been postulated by this author that the fundamental cause of cosmic evolution is the unstable relationship existing between the Anti-G and G forces and that these are also fundamental to space-time geometry. If this assumption is correct, then it must be assumed that the evolution of any other coexisting universes, which one might imagine, would also be driven by this inherent unstable relationship. Regardless of the number of strata that any of these universes may possess, the mathematical structure associated with it (which guides its subsequent evolutionary development) will be determined by the fundamental relationship existing between its oppositional G forces.

This means that all mathematical structures need not exist and that mathematics is merely the symbolic language that logically describes a specific developmental relationship that is contextual and relational in nature. There cannot be an infinite number of real (as opposed to abstract) mathematical structures existing in infinite space, unless there are a correspondingly infinite number of coexisting variations of discrete volume size present. This suggests that unless there is a infinite regional variation in the relationship between the Anti-G and G forces, there cannot be an infinite number of coexisting universes. In addition, neither variation in discrete volume size or in the relationship between the oppositional G forces precludes the possibility of a relativistically induced interaction. Of course, this is only true if the hierarchical stratification model that has been proposed is correct. If it is correct, unless discrete volume size in adjacent regions varies significantly from that of our tristratum universe, relativistic interaction is probably inevitable. This could also be true in spite of variation in each universe's derived forces or parameters.

Such interaction could lead to the intermixture of material from two adjoining universes. This means that when collapse of stratum structure containing information from these interacting universes occurs, it will influence the evolution and development of both in each universe's ensuing cycle. Rather than retaining continued regional variation, it would seem more likely that *universal* uniformity of structure at the most fundamental level would ensure greater cosmic stability. Ultimately, the universal and uniform quantization of both the Anti-G and G forces throughout infinite space ensures the *interdependent* evolution of widely separated regions. The cosmological principle would not be valid if this were not true. If the oppositional G forces operate throughout infinite space, it is difficult to imagine how interaction between distinct universes could be prevented. Even minimal periphery interaction could introduce destabilizing anomalies within our universe, which would impact its homogeneity and invalidate the cosmological principle, or even worse, cosmic collapse in an adjacent region may induce a cascading chain reaction of cosmic collapse, hastening the demise of the universe we call home.

One of the major advantages that an evolutionary cyclic model has over multiverse theories is that it provides an *autocausal explanation* of the apparent fine-tuning of cosmic parameters. In this model, the extracosmic *blind probability* explanation provided by the multi-universe hypothesis is abandoned and replaced by autocausal mutagenetic and selection processes, which are thermodynamically induced. However, one of the roadblocks to developing a successful cyclic model has been the second law of thermodynamics. Models have been proposed that postulate that succeeding cycles in the cyclic process have longer lifecycles and expand with a corresponding increase in maximum size (Y.B. Zeldovich and I.D. Novikov). These models attribute these changes to a cumulative transcyclic transfer of entropy that increases the total entropy existing during initial conditions in succeeding cycles. This can be characterized as the *entropy problem* that has plagued many previous attempts at developing a viable cyclic model.

Paul Steinhardt has proposed a novel approach to this problem through the application of brane theory (an offshoot of string theory). Steinhardt's approach involves an attempt to minimize the impact of increasing entropy density through a process of dilution. This is accomplished by relegating the entropy created during the cyclic process to the branes, which do not appreciably contract during the contraction phase. Instead, it is the curled-up extra dimensions that are contracted during the collision of the branes that initiates cosmic rebirth. During this collision, the older entropy density is rendered irrelevant through its dilution on the branes and is overwhelmed by the entropy density of the new cycle.

Steinhardt's model attempts to address the argument made by Richard Tolman (and others) that there is an increase in entropy density in succeeding cosmic cycles. This is manifested as a cumulative concentration of both older and newer entropy at the big crunch following the contraction of all of the existing space-time dimensions (Steinhardt). It is this increased entropy density that is supposed to be the driving force in the production of a *bigger bounce* in succeeding cycles in accordance with Einstein's theory of general relativity (Steinhardt). As laudable as the Steinhardt approach is in attempting to resurrect the cyclic model, this author does not believe that it addresses the real problem.

Does the second law of thermodynamics have a legitimate *transcyclic application*, or does it have instead a more limited functional role in the cyclic evolutionary process? After all, it is difficult to visualize how the *transcyclic accumulation of thermodynamic refuse* has any natural selection benefit, which would alleviate the fundamentally *inherent instability* that drives cyclic evolution. In addition, if this *transcyclic accumulation* actually occurs, it would suggest that there is increasingly less useful energy available from one cycle to the next. The development of complex structure requires the efficient utilization of useful energy. Any transcyclic depletion of this resource would result in a *regressive evolutionary* tendency toward greater simplicity of structure.

However, nature abhors a gradient because it also *abhors instability*. Therefore, the selection process and the laws of thermodynamics favor the development of complex structure to produce greater efficiency of gradient reduction. A cyclic model that allows for the transcyclic accumulation and the transfer of entropy would produce increasingly inefficient systems of gradient reduction. This would certainly follow if the 2nd Law has absolute transcyclic transcendency, or even if one allows that, it only represents a strong statistical tendency in nature.

Nevertheless, the 2nd Law of thermodynamics alone is insufficient in itself to account for the development of complexity. The ubiquitous presence of open systems in nature suggests that another law, the Law of Maximum Entropy Production, is equally important in explaining thermodynamic phenomena, as well as the arrow of time. The 2nd Law is quite capable of explaining death and disorder, but by itself, it is incapable of explaining a host of phenomena including life itself. With regard to the arrow of time, no chronicler of history focuses solely on a narration of the demise of cultures and civilizations, nor is devolution or extinction the primary focus of the evolutionary biologist (or the cosmologist for that matter). For both the historian and the biologist, the arrow of time represents the exposition of the description, the development, and the succession of complex organization and events. Moreover, the 2nd Law is only part of the reason why one never sees the *cosmic movie* run in reverse. The Law of Maximum Entropy Production is responsible for supplying the characters, the props, and the storyline that makes this cosmic theatrical production interesting, or to use an example, the 2nd Law explains why one never sees a broken egg revert to its previously unbroken state, but it requires the LMEP to explain the origin of the egg's complex structure. Therefore, it requires the interdependent operation of both the 2nd Law and the LMEP to explain time's arrow, or put another way, the arrow of time is demonstrated to be a perception of the complex interdependent interaction of the *incommensurable rivers* of order and disorder (Rod Swenson). It is manifested through the complex interaction of a multitude of interdependent gradient reduction systems.

Open systems reduce their internal entropy by exporting entropy to their surroundings (Schneider and Sagan). In accordance with the LMEP, order is more efficient at producing entropy than is disorder. The LMEP does not violate the 2nd Law but is supplementary to it and compensates for its explanatory limitations. It does this by accounting for the otherwise unexplainable and ubiquitous presence of order in the midst of disorder. Furthermore, it also renders the Boltzmann interpretation, which suggests an infinite improbability for the existence of ordered states, obsolete. The LMEP ensures that ordered states are not only probable but that their existence has a probability of 1 and arise spontaneously whenever critical thresholds are reached (Rod Swenson). The Bénard experiment provides an excellent demonstration of how order can arise spontaneously from disorder when a threshold is attained and when supplied with an energy potential to be minimized.

In the case of the universe, the energy potential is produced via the interaction of the Anti-G and G forces within the discrete units of space before the initiation of cosmic expansion. This takes place throughout infinite space, and the potential is minimized during the expansion phase through the production of *dissipative structure* (Ilya Prigogine). During black hole formation, the Anti-G and G forces are brought into a state of *gravistatic equilibrium* in regions of space where collapse of the stratified space-time geometry has taken place. In these black hole regions, complete dissipation has occurred, and entropy is exported into the surrounding dark energy stratum. The continuous infusion of this energy waste product (which will be explained later) fuels the acceleration of the expansion of the dark energy stratum, expediting its eventual collapse. Upon termination of the cosmic cycle with the final collapse of the dark energy stratum, the oppositional G forces are returned to their natural far-from-equilibrium state. This takes place only after complete dissipation has taken place before cosmic collapse. The reinitiation of the far-from-equilibrium relationship between the Anti-G and G forces throughout infinite space reintroduces the instability that results in re-expansion.

According to the big bang theory, the universe began as a singularity, an infinitely dense point encompassing all of space-time and possessing infinite gravity. A singularity may be viewed as an infinite implosion of space-time, in which gravitational kinetic energy is ever increasing. At the same time, there is a corresponding infinite decrease in gravitational potential energy. The flow of the potential in this case is in the direction of the implosion and is never completely reduced because gravitational potential energy never reaches zero. In this situation, there is no violation of the first law because the gravitational energy is transformed from one form to the other in inverse proportion. It is also a scenario in which space is continuous, a view consistent with general relativity, which is not a quantum theory of gravity.

However, if space is instead composed of an infinite number of discrete units, the boundaries of which are established by the interaction of the oppositional G forces, there can be neither infinite density nor infinite gravity concentrated in a single point. The boundaries of each discrete unit establishes a minimum limit for the operational range of the G force, which also establishes limits on gravitational kinetic and potential energies. If the author is correct, there is an exponential increase of antigravitational kinetic energy compared to that of the G force at the elastic boundaries of each discrete unit of space. Cosmic collapse induces interaction between the oppositional G forces at this elastic boundary, causing an incredibly rapid buildup of pressure. This pressure is relieved when the kinetic energies of both of the oppositional G forces is released through cosmic expansion and the reanimation of the cosmic strata, with a corresponding reactivation of the derived forces.

During the expansion process, some of the kinetic and potential energies of the oppositional G forces are infused into the fields of the derived forces. This effectively diminishes the amount of both forms of energy remaining to the oppositional G forces during the course of the expansion phase. Moreover, this partially explains the relative weakness of the G force compared with the derived forces, as a portion of the field strength of the oppositional G forces is transferred to the derived fields. Once cosmic expansion has been initiated, the stratification of the space-time geometry and the activity of the interacting fields permit the development of complex dissipative structures. These structures are produced to dissipate the kinetic energy released during the far-from-equilibrium interaction of the oppositional G forces at the moment of initial expansion.

Ultimately, the existence of oppositional G forces explains the existence of the cosmic gradient, as well as the function of the 2nd Law of thermodynamics. The 2nd Law combined with the LMEP explains how the cosmic gradient is reduced. Therefore, since cosmic gradient reduction is restricted to the expansion phase of the cyclic process, it appears reasonable to assume that there is a restricted operational role for both the 2nd Law and the LMEP. This does not mean that both laws are irrelevant in the evolutionary processes involved in parametric mutagenesis or mitotic stratification. After all, both have a secondary role in influencing the energy recycling process, which leads to an increase in the thermodynamic pressures that induce cyclic, evolutionary change. However, to reiterate, this is not accomplished through the *transcyclic accumulation of thermodynamic refuse.*

It is precisely because the oppositional G forces possess different ranges and strengths that they cannot ever coexist in a permanent state of equilibrium. Gravistatic equilibrium is only temporarily (relatively speaking) attained during black hole formation because of an inequitable concentration of collapsed stratum structure in local regions of space. However, this is inevitably disturbed upon the final collapse of the remaining majority of the dark energy stratum at the termination of the cosmic cycle. It is the inherently unstable relationship existing between the oppositional G forces that produces the initial conditions that drive the extension of space-time dimensions. The universe reduces its temperature and pressure gradients through expansion and the adaptive processes that produce increasingly efficient dissipative systems of structure. For example, the development of increasingly complex ecosystems more efficiently reduces the solar gradient. In a similar manner, increasing complexity in the evolution of the hierarchically stratified space-time geometry more effectively reduces the cosmic gradient as well.

Ultimately, the universe is on a hopeless *quest* for equilibrium. Unfortunately, this desired state of equilibrium can only be attained locally and temporarily via the production of structures that delay the unstable interaction of the oppositional G forces. The gradual evolutionary production of the derived forces, as well as the hierarchical stratification of space-time geometry, has steadily increased cosmic lifespan. In turn, this has led to a steady, incremental increase in the period of time existing between induced unstable interactions of the Anti-G and G forces. With further adaptation, this trend will continue as the universe increases its dimensional structure in four-dimensional increments with the addition of new strata.

Hopefully, a proper understanding of this process, as well as a mathematical confirmation of the existence of the Anti-G force, will provide solutions to current cosmological puzzles. Perhaps, this will also relegate such ideas as the *many worlds interpretation* (Hugh Everett) of quantum mechanics and the anthropic principle to the dust bin of history. In addition, new insights may be developed to explain some of the peculiarities associated with quantum behavior by exposing the *underlying reality*, the existence of which Einstein suspected. It may be that a revised application of special relativity, which accounts for interstratum interaction, will shed new light on such phenomena as *quantum entanglement* and *quantum tunneling*.

Ultimately, the properties of all *real* particles (*virtual photons* will be dealt with later) are determined by the motion of these particles through the framework of the hierarchically stratified space-time geometry. For example, the differences in angular momentum or spin between fermions and bosons are determined by the manner in which these particles are accelerated (or how they oscillate) through the *twelve-dimensional architecture*. All particles can be viewed as standing waves and harmonic oscillators, with complex variation in their wave patterns as they oscillate through the stratified structure of space-time. Each stratum has its own value for constant c, as well as its own value for the Planck time, and this stratum-specific variation is an example of *hidden variables* that affect the momentum and the position of particles. As a particle oscillates through each stratum of four-dimensional space-time geometry, there is a change in its position and momentum. With each stratum, there is a corresponding variation in motion through time and space in accordance with special relativity. This means there will be variation in time dilation and space contraction, which is stratum-specific. Each particle that inhabits the baryonic stratum is a *standing wave* that is reflected back and forth through the tristratum structure. In contrast, dark matter particles are normally only reflected back and forth through the dark matter and dark energy strata (this is not always true as I will point out when I discuss nucleosynthesis later in this work). For example, as an electron (a spin-½ particle) reaches its peak energy (a crest) during oscillation into its uppermost stratum; it is then reflected back through the lower strata (by the attractive component of the combined offspring forces of the tristratum structure of space-time). It is stopped from further contraction by the combined repulsive force of the offspring forces of the tristratum structure. It will then be reflected back until it reaches peak energy (another crest) again, and this represents one fundamental wavelength. In contrast, spin-1 particles like the photon or W particle experience a more complicated motion through the tristratum structure, which is dependent on the interaction between the dark matter and baryonic matter strata from which they are produced (an attempt at explanation will be made when I discuss nucleosynthesis later on). Unlike a typical standing wave, those that represent particles undergo changes that are determined by the variation in stratum-specific parameters as the wave of energy (of the particle) is reflected back and forth through the tristratum structure.

Because every real particle is constantly in dynamic motion through the tristratum structure as it travels, its actual position and momentum are constantly changing. This may offer a possible explanation of the wavelike nature of particles. The rapid oscillation of these through the tristratum structure means that they are only briefly present in each stratum, which is why a wave function can only make probabilistic predictions unless there is observer interaction. Higher energy particles oscillate more rapidly through the tristratum structure and are subjected to greater time dilation than lower-energy particles. In addition, as a particle oscillates through each stratum, its change in energy occurs at Planck time intervals that are stratum-specific. Higher-energy particles have greater changes in energy with each Planck time interval than lower-energy particles.

A rough way to visualize a particle like an electron is as a *spinning ball of energy*, which is oscillating back and forth with angular momentum through the tristratum structure of space-time, or if one chooses a string theory interpretation, what one might visualize as a *spinning ball of energy* is instead a tiny string, which is vibrating within four-dimensional space-time when within the baryonic stratum. From the perspective of an observer in the baryonic stratum, it (the particle or vibrating string) will appear to materialize and dematerialize repeatedly at a rate dependent on its energy. This can be represented as stratum-specific changes in the x, y, z, and t coordinates as the particle (or vibrating string) undergoes materialization and dematerialization during oscillation through the tristratum structure. Also, from the observer's perspective (as viewed within the baryonic stratum), the wavelength of a particle is the distance in meters (or fraction of a meter) between successive peak materializations of the particle within the baryonic stratum, and the frequency is the number of materialization/dematerialization cycles in a given unit of time. Therefore, the observed frequency and wavelength are dependent on both the interstratum (tristratum oscillatory motion) and the intrastratum momentum of any particle.

Measurements made on particles require an interaction with them, which impacts their motion through the tristratum structure. Anytime an experimenter attempts to locate the position of a particle with another particle, for example, a photon of high frequency and energy, there will be an exchange of momentum and energy. This, in turn, affects the motion of the measured particle through the tristratum fabric, as well as within the baryonic stratum. The transfer of energy and momentum from the incident photon will increase the rate of oscillation of the measured particle through the tristratum fabric. Because the value of the constant c is increasingly greater, descending downward from the baryonic to the dark energy stratum, there will be a corresponding variation in time dilation. This is in accordance with a stratum-specific application of special relativity.

The oscillating particle (or vibrating string) will experience the greatest time dilation while it oscillates through the baryonic stratum. It will also experience time dilation while it oscillates through the *substrata* but to an increasingly lesser degree (in accordance with the stratum-specific variation of the constant c operating in these strata). With increasing energy, time dilation effects as observed when the particle has materialized in the baryonic stratum become more and more pronounced. This results in relatively longer duration of materialization within the baryonic matter stratum; hence, it allows for a more precise determination of the position of a particle. In addition, according to special relativity, contraction of space occurs in the direction of motion. Since one direction of motion is through the tristratum structure, an increase in the rate of oscillation will result in an overall contraction of the tristratum structure at the location of the oscillating particle. The degree of space contraction in the direction of oscillation will also vary from stratum to stratum. As a result of the combined tristratum effects of time dilation and space contraction, the observer is able to determine the location of the particle with increasing precision as its wave packet (and the tristratum structure it oscillates within) is compressed. The degree of precision will correspond to the amount of change there is in the oscillation rate of the measured particle. The uncertainty principle indicates that this achieved greater precision in the measurement of position is at the expense of increasing imprecision of the particles' momentum.

Only a portion of the momentum from an incident particle may be reflected in the change in oscillation rate of the measured particle. The remainder may impact the momentum of the particle, as it is normally observed traveling within the baryonic stratum (for particles that inhabit the baryonic stratum). According to Heisenberg, if the observer has more precise knowledge of the particle's position, he/she now has correspondingly less information about its momentum (and its wavelength) before measurement. All like particles (electrons for example) with identical energy will dilate time and contract space in an identical manner as they oscillate through the tristratum structure. If the rest mass and oscillation rate (while at rest) of a given particle, like an electron with minimal energy, can be precisely determined (if it is universal), its momentum in the direction of oscillation through the tristratum structure can also be determined. At minimal energy (and minimal momentum) with correspondingly very little dilation of time, a particle like an electron will have an ill-defined position because it is oscillating through the strata at a relatively equal amount of time. However, at higher energies, the disparity in time dilation becomes more pronounced, and the particle would appear to be in the uppermost stratum with peak energy for a longer period of time during oscillation. Its position is no longer ill-defined from the observer's perspective, and with increasing energy, time dilation increases, and the observed position of the particle becomes increasingly more precise. It is clear that changes in time dilation are dependent on changes in interstratum momentum (as well as intrastratum momentum). If the variation in time dilation of an energetic particle from a standard momentum of the same particle at rest can be determined, then a precise measurement of the change in total momentum (as a combination of interstratum and intrastratum momentum) can be determined as well, at least in principle. This applies to photons as well as it does for other incident particles. In other words, if the position of a particle is precisely determined, then it does not possess an indefinite wavelength (and indefinite momentum), as Heisenberg proposed. The problem that ultimately complicates measurements is that a particle is undergoing continual stratum-specific fluctuations in both position and momentum as it oscillates through the hierarchically stratified space-time structure. Changes in energy due to observer interaction introduce relativistic effects that are stratum-specific and must be accounted for when determining respective changes in position and momentum.

For the photon, its momentum and energy are dependent on its frequency (or equivalently, its wavelength), and this frequency is, in turn, a measurement of the oscillation rate of the photon through the tristratum structure of space-time (since photons are massless and always travel at the speed of c). A faster oscillation rate means a greater energy, a shorter wavelength, and a correspondingly greater momentum for the photon in the direction of oscillation through the tristratum structure. Higher-energy photons will materialize within the baryonic stratum at a more rapid rate than do lower-energy photons. However, since photons always travel at the velocity c within the baryonic stratum, and c is the parameter that determines time dilation as measured within the baryonic matter stratum, a problem arises. For particles with mass, time dilation effects arise as a consequence of both intrastratum and interstratum momentum. It is these time dilation effects that make a position determination possible for fermions. If a photon is subject to time dilation effects, then it could not always be traveling at the intrastratum velocity c (as measured within the baryonic stratum). In fact, its velocity would vary depending on the energy of the photon because of corresponding time dilation effects (the higher the frequency, the slower the photon would appear to travel). (Time dilation effects are minimized during oscillation by a stratum-dependent variation in interstratum momentum as photons oscillate through the lower strata. However, they continue to travel at the velocity c as measured within the baryonic stratum during intrastratum materialization in these lower strata.) Clearly, this is not allowed; therefore, photons cannot undergo time dilation effects as they materialize within the baryonic stratum of the tristratum structure of space-time. On the other hand, the weak interaction particles (W+, W−, and Z) do possess mass, and because of this, there is some time dilation as they materialize in the baryonic stratum before they decay.

Although photons do not experience time dilation effects during materialization within the baryonic matter stratum, the photon's frequency (related to its rapidity of oscillation through the tri-stratum structure) does impact the rate of materialization and dematerialization that it undergoes during travel. All optical phenomenon, including refraction, reflection, diffraction, and interference, arise as a consequence of the interstratum oscillatory motion of photons. In addition, the experimental results of the famous single—and double-slit experiments can be readily explained in terms of this new understanding of quantum behavior. It is these experiments that are used to demonstrate the *wave/particle duality* of particles like the photon and the electron (for which the experimental apparatus is somewhat different, but the results are the same). Included in the setup is a point source from which photons are fired at a barrier with two slits, which eventually emerge on a screen at a set distance from the two slit barrier. When the right slit is covered the emerging photons form a simple *diffraction pattern* on the left side of the screen, much like bullets. If the left slit is covered, then the emerging photons form a similar pattern displaced to the right side of the screen. It is this single-slit variation of the experiment that illustrates the particle aspect of photons. However, when both slits are left open, instead of forming two bands representing a simple diffraction pattern, there will instead appear an *interference pattern* of separated fringes. This result is said to illustrate the *wavelike nature* of the photon. What are we to make of this apparent paradox? Current theories offer no satisfactory explanation, but that is because they do not account for the complex stratified, structure of space-time and its role in explaining quantum behavior.

As mentioned earlier, a particle can be roughly visualized as a *spinning ball* of energy (or vibrating string), which is oscillating through the tristratum structure, alternately materializing and dematerializing within each stratum (each with its own four-dimensional space-time structure) during this oscillatory motion. In this sense, we can roughly visualize a photon alternately *appearing and disappearing* as it travels at the speed of light from the point source through the barrier and eventually striking the screen. The intrastratum wavelength of the photon as measured within the baryonic layer will depend on the rate (or frequency) at which the photon materializes during its intrastratum travel. In the single-slit experiment, there is minimal interaction between the photon and the slit as it passes through, and the combination of spin, polarity, deflection angle (or scattering angle), and the state of materialization during interaction with the slit will determine where it strikes the screen. In short, the photons will accumulate on the screen in a pattern very much like bullets would.

In the two-slit experiment, the photon (or the vibrating string that represents it) is essentially *smeared out* by interaction with the two slits during intrastratum materialization. Because a string is an extended object, which is vibrating, parts of it will be materializing at different rates, which will impact its direction of motion. (The intrastratum wavelength of the photons cannot be too long or the photons will not be in a state of materialization within the baryonic stratum when they pass through the slits.) Where each photon strikes in this case will also depend on spin and polarity as in the single-slit case. However, the difference is that the materializing photon is smeared out (or the vibrating string undergoes distortion), and the extent of this smearing will depend on the slit width and both the rate and the state of materialization during interaction with the two slits. Photons with longer wavelengths will have a slower rate of materialization and are more likely to vary in their state of materialization during interaction with the two slits, resulting in a larger average scattering angle upon emergence from the barrier. In contrast, shorter-wavelength photons will have a faster rate of materialization and a smaller, average scattering angle. In both the single—and double-slit experiments, the photon is oscillating through the tristratum structure (its wavelike behavior), and when it strikes the screen, it interacts with the screen like a particle (an intrastratum interaction). This is because interaction between the photon and an atom can only occur when both the photon and an electron in one of the screen's atoms is materialized within the baryonic stratum.

The interstratum and intrastratum motion of electrons is also important in explaining atomic structure. Electrons do not orbit the nucleus in quite the same manner that planets orbit the sun. This is a consequence of the interstratum oscillation of these particles, which results in materialization and dematerialization to occur within the various strata as they move in space-time. The electron will trace a near-circular path with varying eccentricity and orbital inclination, about the nucleus (with orbital distance and dependent variation in angular velocity). During this motion, an electron will continuously materialize and dematerialize at a rate (frequency) dependent on the *orbital* it occupies. An observer can only locate it when it has materialized within the baryonic stratum. However, a portion of the periodic revolution about the nucleus will involve motion during materialization within the lower strata, during which the electron cannot be located by an observer. The region in which the electron can materialize within the baryonic stratum will depend on a number of factors, including its relationship to other electrons in their various orbitals and its interaction with the nucleus.

Electromagnetic repulsion is one factor in determining where a given electron will materialize, and this is mitigated, in part, through variation in the interstratum oscillation rate (which is orbital and distance dependent) of each particle. It is also alleviated through variation of the timing and location of materialization within the baryonic stratum of both electrons that inhabit the same orbital, as well as those which inhabit nearby orbitals. For example, within the 1s orbital, one electron may be fully materialized within the baryonic stratum, while the second is oscillating through the lower strata (the same holds for each of the *p*, *d*, and *f* orbitals as well). The shape of the various orbitals represents the restricted regions of space where an electron may materialize within the baryonic stratum. Operation of the electromagnetic force is *stratum restricted*, as is the electroweak force that operates only within the dark matter stratum. This means that electrons that are in a *state of materialization* within the baryonic stratum exert no electrostatic repulsion on electrons that are simultaneously oscillating through the lower strata and vice versa. The electroweak force operating within the dark matter stratum does exert a repulsive force between electrons that are simultaneously oscillating through the dark matter stratum, but this force is weaker than the baryonic stratum's electromagnetic force.

Because of this stratum-dependent restriction of force influence, it is possible for an electron to pursue an elliptical path of orbit about the nucleus. However, this is not representative of the orbital shape, which only depicts the region in which it may materialize within the baryonic stratum. This means that a p orbital electron, like its s orbital counterpart, may also trace a roughly circular (with varying eccentricity) orbit about the nucleus but will be oscillating through the lower strata during a portion of its orbital travel. The characteristic dumbbell shape of the p orbitals just represents the restricted region in which p orbital electrons may materialize within the baryonic stratum.

The shape of the earth's orbit about the sun (which is situated at one foci of an ellipse) changes periodically because of the gravitational influence of the other planets in our solar system. These planetary influences introduce changes in the elliptical eccentricity, the orbital inclination, and the apsidal precession of the earth's orbit. For the most part, the masses of the earth and the sun remain relatively constant (if one ignores the loss of mass due to nuclear fusion processes); therefore, the *barycenter* would remain constant without multibody influence. However, the planets and the sun are macroscopic objects, the motions of which are measured within the baryonic stratum. This is despite the fact that they are massive collections of particles that are oscillating through the tristratum structure. It is because the particles that compose them are undergoing differing rates of materialization and an observer only measures their collective behavior as measured within the baryonic stratum that classical laws are applicable. For the most part, one must only consider the forces and the parameters (the gravitational constant G, the constant c, etc.) that hold sway within the baryonic stratum, and that is why classical laws work. Within the atom, every one of its bodies, including the electrons, the protons, and the neutrons, is subject to variation in forces and parameters, which are stratum dependent, as each of these oscillates through the tristratum structure.

Unlike the sun and the planets, the motion of the electrons are not dependent on gravitational forces but by the collective influence of the electromagnetic force within the baryonic stratum, the electroweak force within the dark matter stratum, and the strong-electroweak force of the dark energy stratum. However, because the operation of these forces are restricted to each respective stratum, only those electrons, protons, and neutrons that are materialized within a given stratum at the same time are affected by that stratum's derivative forces. Like the electrons, the protons and the neutrons are also oscillating through the tristratum structure as collections of quarks and gluons. Each proton is composed of two up quarks and one down quark, and each neutron is composed of two down quarks and one up quark. Color charge is a property of the quarks combined within these protons and neutrons. Each nucleon must have a combination of red, green, and blue charge to produce a neutral color (not to be confused with actual colors in the normal sense), and the relative combination of these may change as the nucleon oscillates through the tristratum structure. What causes a green up quark to change to a red up quark during rematerialization within the baryonic stratum arises as a consequence of the fact that the gluons, which hold the quarks together, are also undergoing oscillation at their own rates through the tristratum structure.

Although the three quarks that comprise a proton or neutron may collectively undergo oscillation at the same time, all of the nucleons within the nucleus will be oscillating out of sync with each other. This means that different nucleons will be materializing at a given instant within the baryonic stratum, while others are materializing within the lower strata. One important effect of this variation is a fluctuating change in barycenter between the nucleus and the electrons that orbit it. It is the complicated relationship between the oscillating nucleons and the oscillating electrons that effect changes in barycenter and other variations that shape the individual orbits of each individual electron. It is all of these factors, along with the stratum-dependent variation in force strength and velocity of propagation, that will determine where a given electron will actually materialize within a given orbital at any one instant. Therefore, it should be possible in theory to formulate a quantum mechanical variation of Kepler's laws that incorporates the tristratum oscillatory motion of subatomic particles, as well as the stratum-dependent variations of the repulsive and attractive forces, which also influence orbital motion.

Another important point that needs to be addressed at this time is the orbital dependent variation in frequency of the electrons. The electrons that inhabit orbitals at increasingly greater distances from the nucleus will have higher frequencies and energy than those that are nearer to it. These outer shell electrons will be oscillating at a faster rate (which is another factor in limiting where they may materialize within the baryonic stratum) through the tristratum structure and therefore be subject to greater time dilation as they materialize within the baryonic stratum. This means that the more-distant (from the nucleus) orbital electrons will behave more particlelike, and this partially explains (among other reasons) why it is only the electrons inhabiting the outermost orbital (the valence electrons) that are involved in chemical reactions.

The stratification of space-time structure explains another phenomenon, which has continued to perplex the physics community. *Quantum entanglement*, the so-called spooky action at a distance, has a possible explanation that arises as a consequence of stratum-dependent variation in parameters (specifically the value for the constant c). From the perspective of an observer in either the dark matter or dark energy stratum, photons that exist within the baryonic stratum travel at a velocity slower than the stratum-dependent velocity of c operating within the substrata. This means that from an observer perspective in these lower strata, photons can be treated as though they have inertial and gravitational mass. In addition, events that occur outside of *light cones* (past and future) within the baryonic stratum may occur within the light cones of the substrata. Because entangled particles are causally connected as they oscillate through the tristratum structure, it is possible for information to be exchanged between them at a velocity faster than the speed of light, as it is measured within the baryonic stratum.

As a consequence of the stratum-dependent variation in the constant c, even entangled photons can exchange information between them at a velocity that exceeds their own, as measured from the perspective of an observer in the baryonic stratum. This means that the Lorentz equation for addition of velocities ($V = v1 + v2 / 1 + v1v2 / c2$) can be applied in a stratum-dependent manner, to photons, as well as other particles. One only needs to know the values for c, which operates within the lower strata, to determine how quickly information can be exchanged.

In addition, because photons have inertial and gravitational mass from a substrata perspective, both dark matter and dark energy may interact gravitationally with them. (It is only the derivative or offspring forces that are subject to stratum-dependent restriction of operation, with exceptions that will be explained later in this work.) This is an explanation for the gravitational lensing of light by the halos of dark matter that enshroud galaxies. In addition, the bending of light around massive bodies can be explained by the concentration of dark matter that contributes to their structure. Although the gravitational influence of dark energy is relatively weak because of its isotropic distribution, its primary effect is to lengthen the wavelength of the cosmic background radiation during cosmic expansion. In addition, its isotropic distribution (except primarily where a portion of the dark energy stratum is collapsed within black holes) permits photons to travel in a straight line path in otherwise empty space.

Furthermore, because dark energy particles only oscillate within the dark energy stratum, their primary gravitational influence on a photon is during the brief period during tristratum oscillation when the photon is materializing within the lowest stratum. This impacts the interstratum oscillation rate of the photon (and not its intrastratum velocity), which means it will materialize with decreasing frequency during its travel at the velocity of c, as measured in the baryonic stratum. Of course, because of the weakness of this interaction, it took an incredibly long time (billions of years) for the wavelength of the photons produced shortly after the big bang to become that which represents the current microwave background radiation. It is increasingly clear that cosmological models will have to be modified and take into account the complex hierarchical structure of the space-time geometry. This means that many of the currently accepted ideas will be discarded, but at the same time, new explanations will be found for phenomena that are not presently understood.

However, before I proceed further in this direction, I would like to attempt to explain virtual photons (as promised earlier), within the context of the tristratum space-time framework. According to theory, the electromagnetic force is communicated between electrons and protons via the exchange of virtual photons. Real photons, although massless, possess momentum and energy, which is related to an individual photon's frequency. In addition, each of these travels with the velocity c, which is approximately 299,792 kilometers per second in a vacuum. Of course, this is only the value for c as measured within the baryonic stratum. Virtual photons as force carrier particles have only a temporary existence and can never be detected (although their effects can be observed).

Keep in mind that the value for the constant c within the dark matter and dark energy strata is greater than 300,000 kilometers per second. As mentioned earlier, an intrastratum signal exchanged between two entangled particles, including real photons, travels at the velocities of c as measured within these lower substrata. A virtual photon traveling at these intra-stratum velocities will not be materialized within the baryonic stratum long enough for it to be detectable (which preserves conservation of energy). During oscillation through the lower strata, a virtual photon has a stratum-dependent variation in its intrastratum velocity as it oscillates through these lower strata. This means that when a virtual photon is exchanged between two particles, it will be traveling faster than 299,792 kilometers per second during its travel when it is materializing within the lower strata.

On the other hand, real photons do not vary their intrastratum velocity as they oscillate through the tristratum structure (they always travel at 299,792 kilometers per second). In addition, real photons may only interact with particles via intrastratum interaction within the baryonic stratum. In other words, both interacting particles, the photon and an electron (or proton), must be materializing within the baryonic stratum for interaction to occur. In contrast, a virtual photon may be exchanged between an orbiting electron and a proton within a nucleus during the simultaneous materialization of these particles within any of the three strata. This permits continuous interaction between orbiting electrons and the nucleus. For example, an electron and proton simultaneously materializing within the dark matter stratum will only receive virtual photons, which are also materializing within that same stratum. Moreover, this exchange will occur when the photon is traveling at the stratum-dependent velocity operating within that stratum and with a corresponding stratum-dependent variation in force strength. However, as pointed out earlier, this same photon traveled at a stratum-dependent variation in velocity during interstratum oscillation between the time it was emitted and the time it was absorbed.

There is also a distance-dependent variation in interstratum momentum for virtual photons exchanged between the protons and the orbiting electrons. Interstratum momentum (for the virtual photons) increases at increasingly shorter distances and decreases at increasingly greater distances. It is this distance-dependent variation in interstratum momentum that explains the distance-dependent variation in force strength. This variation in interstratum momentum does not impact the intrastratum velocity of the virtual photon as it oscillates through the tristratum structure. However, it does result in a variation in intrastratum transfer of momentum (depending on distance) between the protons and the orbiting electrons, which in this case, corresponds to an increase in attraction at shorter distances (and a distance-dependent variation in the repulsive force between electrons). (There may be a possible minimum limit to interstratum momentum for virtual photons. This may suggest that there is a limit to the range of the electromagnetic force corresponding to this minimum limit; in contrast, real photons can travel to tremendous distances, though perhaps not infinite distances as will be explained later.) The complex orbital structure within atoms depends on this distance-dependent variation in interstratum momentum, as well as the intricate variation in stratum-dependent variation in exchange velocity, force strength, and intrastratum restriction on exchange.

Another type of virtual particle that is hypothesized to be the exchange particle involved in the gravitational force is the boson referred to as the *graviton*. Like the virtual photon, the virtual graviton also travels at a varying intrastratum velocity of c during interstratum oscillation when it is exchanged between objects with mass. This means that when it materializes within the lower strata during interstratum oscillation, it will possess an intrastratum velocity that exceeds that of c, as is measured within the baryonic stratum (it will be traveling at the velocity for c associated with the lower strata). In addition, like the virtual photon, a virtual graviton may interact with an oscillating real particle within any of the strata that it oscillates. Virtual gravitons associated with the baryonic matter stratum will oscillate through the entire tristratum structure

DIAGRAMS OF VIRTUAL PHOTON EXCHANGE BETWEEN A NUCLEUS AND AN ORBITING ELECTRON

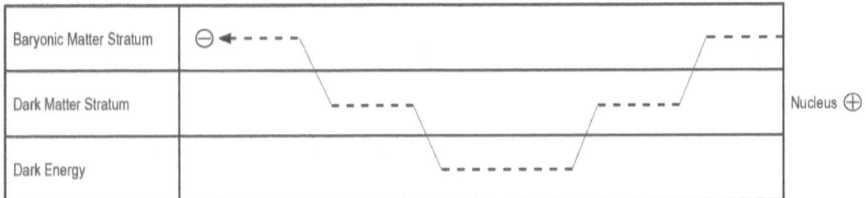

The exchange is faster than the constant c as measured in the baryonic matter stratum

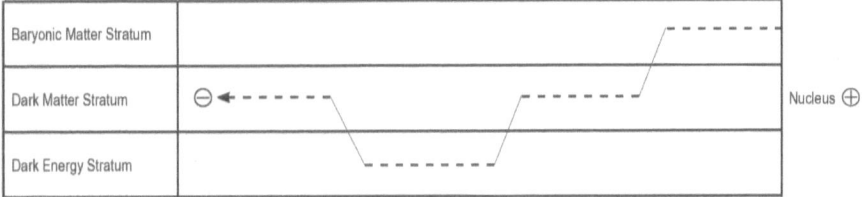

The exchange is faster than the constant c as measured in the baryonic matter stratum

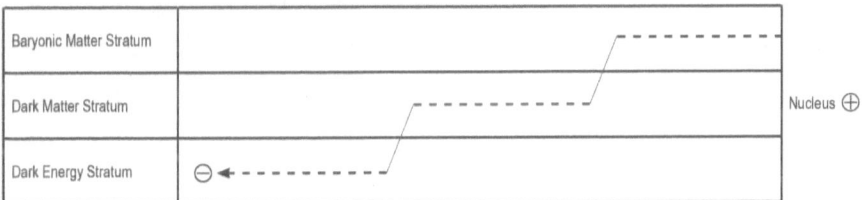

The exchange is faster than the constant c as measured in the baryonic matter stratum

An electron is oscillating through the tri-stratum structure as it orbits a nucleus. Dotted lines represent the travel of a virtual photon as it is exchanged between the nucleus and an orbiting electron (as it appears in each stratum during oscillation through the tri-stratum structure). Virtual photons travel faster than real photons because they change velocity in accordance with stratum dependent variations in the constant c as they oscillate through each stratum (c is correspondingly faster in the lower strata). Real photons oscillate through the tri-stratum structure but they do not change their velocity as they oscillate through each stratum. Note: the exchange velocity pattern of virtual gravitons is also subject to stratum dependent variation in the constant c. Dark energy constant c > dark matter constant c > baryonic matter constant c.

of twelve-dimensional space-time geometry. In contrast, virtual gravitons exchanged between dark matter will oscillate through the two lower strata with its eight-dimensional geometry, and dark energy gravitons will be limited to exchange within the dark energy stratum and its four-dimensional space-time structure.

Any virtual graviton may interact with any of the three types of matter, but it is restricted to intrastratum interaction within the strata it is permitted to oscillate. For example, a dark energy graviton may interact with a baryonic matter particle, but it may only do so when the baryonic matter particle is materializing within the dark energy stratum as it oscillates through the tristratum structure. In like manner, a baryonic matter graviton may be exchanged with a dark matter particle when the baryonic matter particle is oscillating within the two lower strata (of eight-dimensional space-time structure). The strength of the G force between the particles will depend partially on the stratum in which intrastratum exchange occurs and its associated value for the constant G. It will also depend on the distance between particles when the interaction occurs. At shorter distances, virtual gravitons will possess greater interstratum momentum, and at increasing distances interstratum momentum will decrease in inverse proportion to the distance squared in accordance with Newton's law. The intrastratum measurement of frequency and wavelength depends on the complex relationship between the intrastratum and interstratum momentum of the particular graviton. The graviton will possess a stratum-dependent variation in wavelength (and frequency), which also depends on the stratum in which intrastratum exchange occurs. The shortest possible wavelength in any stratum can never be smaller than the Planck length because of the presence of the antigravitational force (and this is only attained in the process of gravitational collapse leading to black hole formation).

It is also probable that there is a longest possible wavelength (which is stratum dependent and represents the lowest possible interstratum momentum) a virtual graviton (like the virtual photon) may possess, which suggests that the G force may have a tremendous but limited range. (The shortest possible range of the G force is at a shorter distance than that of the derived forces because the derived forces cease to function at a slightly greater distance than does the G force. This is the state of affairs involved with gravistatic equilibrium within black holes, which establishes a temporary limit, until the cosmic expansion phase ends, on the shortest operational range of the G force until the next big bang. It is the interaction of the G and Anti-G forces within black holes that confines the collapsed fields of the derived forces during gravistatic equilibrium.) Each stratum's derived forces play a role in determining both the interstratum momentum and the wavelength of gravitons by expanding spatial dimensions and acting in opposition to the G force. This is because the derived forces in each stratum are stronger than the G force, except at the shortest distances. It is only strong multistratum interaction of the G force (which requires the presence of concentrations of both dark and baryonic matter) that permits the overpowering of each stratum's derived forces. The constant c associated with each stratum also has a role in establishing the maximum attainable interstratum momentum (the maximum possible interstratum momentum represents the maximum strength of the G force) a virtual graviton may possess. Hopefully, once the stratum-specific parameters of each stratum are determined, it will be possible to develop equations that describe the interstratum and intrastratum momentum of virtual gravitons. Once this is accomplished, it is also hoped that it will then be possible to ascertain the maximum and minimum strengths and ranges of the gravitational force (if such exists). In any case, it is probable that a successful quantum theory of gravity can only be developed when the complexity of space-time structure is fully accounted for.

Of course, the previously unrecognized complexity of the space-time structure is not just limited to explaining quantum behavior. For example, the tristratum interaction of the G and Anti-G forces may explain why there are three families of elementary particles with variations in mass. Normally, these oppositional G forces interact weakly. However, in situations where the tristratum structure is caused to contract, either gravitationally within massive macroscopic bodies or in high-energy collisions, tristratum interaction of the oppositional G forces is increased. For example, the muon is more massive than the electron because of the manner in which it is influenced by the interaction of the oppositional forces as it oscillates through the tristratum structure.

The muon contracts the tristratum structure more than the electron, and this facilitates interstratum interaction of the oppositional G forces. Unfortunately, for the muon (as is the case for the massive variants of the other families), this contraction of the tristratum structure is unstable. The oscillating muon is subject to influence by the derivative forces as it materializes within each stratum. The contraction of the tristratum structure induces interaction between the electroweak force of the dark matter stratum and the weak force of baryonic stratum. Once the cause of the contraction (see above) is removed or terminated, these derivative forces assert their influence, and shortly thereafter, the tristratum structure expands and the muon decays via the weak interaction (into an electron and two neutrinos). Sometimes electron-positron pairs are produced during this decay from a muon to an electron.

As was mentioned earlier, the dark matter stratum is the source of all antimatter, and it is no coincidence that decay products of second—and third-family elementary particles sometimes include antimatter particles. It is the induced interaction between the dark matter and baryonic matter strata that permits an exchange of matter and energy between the two strata. Positrons are a dark matter (which is normally restricted to oscillation within the lower-two substrata) contribution of matter, which is induced to oscillate through the tristratum structure. The variation in parameters, as well as the variation in the number of strata, these two types of matter oscillate, prevents annihilation events under normal circumstances. It is only when the two strata are induced to interact through contraction of the space-time geometry that dark matter is caused to oscillate into the uppermost stratum (baryonic stratum). From a thermodynamic perspective, it is the complex interaction between the fields (of force) of the dark matter and baryonic strata that is responsible for the production of high-quality energy via annihilation events. It is this high-quality energy that is made readily available for use in gradient reduction by the innumerable open systems that exist throughout the universe.

Interstratum interaction is responsible for the process of nucleosynthesis, which takes place within the cores of stars. From the standpoint of classical mechanics, the temperature in the cores of stars like our sun is insufficient to provide the energy needed to overcome the electromagnetic repulsion between protons during nuclear fusion. Current theory proposes that particles that would normally not be capable of doing so have a small probability (due to Heisenberg's uncertainty relation) of overcoming an energy barrier (of the potential well) via quantum tunneling (Gribbin). (I have never been satisfied with this explanation since I first became aware of it. It just seems too contrived to me and somewhat of a cop out.) A better explanation for this so-called quantum tunneling behavior, at least in the case of nuclear fusion, recognizes the gravitational influence of dark matter in providing the necessary additional energy to overcome the Coulomb barrier.

As for the riddle of how stars, which classical mechanics propose, are not hot enough to induce nuclear fusion, there is now an answer. All of the instruments modern science has for measuring temperature are admirably capable of doing so as long as what is being measured primarily inhabits (that is, it oscillates into) the baryonic stratum (only an intrastratum measurement of temperature is possible). However, from the perspective of an observer who lives within the baryonic stratum, dark matter has no measurable temperature because it normally oscillates only through the dark matter and dark energy strata (and its combined eight-dimensional structure: six space dimensions and two time). (There is no direct way to measure the temperature of dark energy either because it oscillates only within the lowermost strata of the four-dimensional space-time geometry.) This means that even though the gravitational influence of dark matter within the core of a star is instrumental in providing the energy to overcome the Coulomb barrier, it is primarily the baryonic matter present that contributes to the core temperature. The combined influence of the gravitational fields of both the dark matter and the baryonic matter concentrated within a star's core contracts the strata, inducing interaction between the derivative fields of these two strata.

This interaction is minimized and controlled owing to hydrostatic equilibrium and the contribution of dark matter (in the form of antiparticles), which is induced to oscillate into the baryonic stratum. The positrons produced in this manner soon annihilate with electrons in the plasma, producing high-quality energy in the form of gamma rays. It is the thermal and radiation pressures resulting from this process that counteract the gravitational influence of the dark matter and the baryonic matter within the core. Each fusion event reduces the bistratum gravitational gradient in this manner, and the core temperature as measured within the baryonic stratum remains relatively constant. Wherever a fusion event takes place (during the hydrogen burning phase) within the star, the induced interaction between the two uppermost strata is temporarily lessened, with the release of a positron (which annihilates with an electron) and a neutrino. Further reactions occur as bistratum interaction is renewed and helium is eventually produced (the proton-proton chain will be discussed in more detail later). The gamma rays produced are repetitively absorbed and emitted, giving up energy as they slowly travel through the layers of the sun. Eventually, some of the photons arrive on planets at varying frequencies (including that of visible light), and their energy becomes available for further thermal gradient reduction. It is clear that interstratum interaction has an important role in reducing the cosmic gradient, as well as explaining quantum behavior. The stratification of the space-time geometry into three strata also explains a host of macroscopic phenomena, including supernovae, the formation of neutron stars, black holes, and galaxy formation.

It is widely accepted that the gravitational influence of dark matter has had a significant role in the formation of macroscopic structure throughout the universe since the time of the big bang.

The accumulation of baryonic matter to form galaxies early on is attributable, at least in part, to this influence. This galaxy formation was accelerated because of the relatively high density of both dark and baryonic matters in the early universe. The combined gravitational interaction of both types of matter hastened the tristratum collapse of space-time structure, which resulted in the formation of the supermassive black holes at the center of the galaxies. There is a contraction of the tristratum fabric in the regions of space in which galaxies form. This allows for an increased gravitational interaction between the two types of matter (dark and baryonic) comprising the galaxies and the local concentration of dark energy with which it interacts. Ordinarily, there is only weak interaction between dark energy and the other two types of matter. However, in regions in which mass concentration is high, the gravitational property of dark energy reduces the influence of its repulsive property. This permits dark energy to accumulate at higher density within massive macroscopic structures. The end result is that cosmic expansion takes place outside these regions of high dark energy density and the distance between galaxy clusters and many galaxies steadily increases. Although the dark energy stratum does not play a significant role in stellar processes in massive stars, combined tristratum interaction results in the collapse of a portion of the dark energy stratum during black hole formation (as mentioned earlier).

In normal stellar processes, the dark matter stratum has a fundamental role in the nuclear reactions that occur within the cores of stars. This stratum's contribution to the gravitational gradient is instrumental in overpowering the Coulomb barrier during nucleosynthesis. An alternative explanation of this phenomenon is now possible, which dispenses with the somewhat arbitrarily devised notion of quantum tunneling. As has already been mentioned, interstratum interaction permits reactions to occur at temperatures that defy conventional understanding. Richard Feynman once said, "I think I can safely say that nobody understands quantum mechanics." The reason for this is that a significant amount (although not all) of the uncertainty that attaches to the uncertainty principle is due to our present ignorance of the hierarchical structuring of the space-time geometry, and how it influences quantum behavior.

One important consequence of this hierarchical structuring is the complex interplay between interstratum gravitational forces and derived forces, which is fundamental to the process of nucleosynthesis. The mutual gravitational attraction between dark matter and baryonic matter permits an accumulation and increase in density concentration of both. If the concentration becomes sufficient, contraction of both strata associated with these two types of matter is initiated. This contraction is then responsible for inducing interaction between the derived forces of these strata, and nuclear reactions commence. It is possible that the neutrinos produced in these early reactions have a role in limiting the size of a star once it begins this process. Although it is believed that neutrinos only weakly interact with normal matter, it is not yet known whether this is true for dark matter. In fact, neutrinos have been suggested as a possible candidate for dark matter (along with WIMPs—**W**eakly **I**nteracting **M**assive **P**articles and MACHOs—**M**assive **A**strophysical **C**ompact **H**alo **O**bjects), but if their production is via bistratum interaction, they may have a more complex function. If at least a portion of solar neutrinos (there are three types of neutrino: (1) electron neutrino, (2) muon neutrino, and (3) tau neutrino) are capable of interacting with dark matter, then there are at least two ways in which they may limit the size and the density of stars.

As has already been mentioned, the photons produced as the result of positron-electron annihilation are instrumental in producing hydrostatic equilibrium. This equilibrium state limits further contraction of a star through fusion processes, and this in turn limits gravitationally induced contraction of the dark matter stratum. As products of matter-antimatter annihilation, the gamma rays produced reduce the gravitational gradient through the export of energy (via mass conversion), which is contributed from both strata (the dark and baryonic matters). However, photons only interact with baryonic matter.

Neutrino production is also the result of interstratum interaction and contributes to gravitational gradient reduction as well. If solar neutrinos are capable of interacting with the dark matter concentrated in the core of a star and its outer layers, they may be responsible for limiting the overall density of dark matter within the star. By limiting the accumulation and the concentration of dark matter within the star, neutrinos also indirectly limit the accumulation of baryonic matter within its boundary. In this manner, the production of both photons and neutrinos contributes to hydrostatic equilibrium between the two strata. Neutrinos may also interact with the local interstellar space surrounding the star, preventing further accretion of dark matter within the intrastellar boundary. Moreover, the continuous production of neutrinos contributes to the maintenance of uniform density of the interstellar dark matter, which limits its gravitational influence. Therefore, it is the complex relationship between interstratum and intrastratum interactions that is responsible for the relative stability of all macroscopic structure. Stellar nuclear reactions are one example of how interstratum interactions produce observable intrastratum processes like hydrostatic equilibrium.

The primary means by which stars with the mass of our sun (as well as those with lesser mass) convert hydrogen into helium is via a process called the proton-proton chain. During the first step in this process, two hydrogen nuclei are fused to form deuterium, an isotope of hydrogen that contains a neutron in its nucleus. This occurs as one of the up quarks within one of the fusing protons is converted into a down quark, transforming the proton into a neutron. The Coulomb barrier is an intrastratum restriction, which is overcome via the combined interstratum reinforcement of the gravitational forces of the dark matter and baryonic matter strata. This results in the contraction of both strata, which induces an interaction between the (relatively weaker) electroweak force of the dark matter stratum and the electromagnetic and weak forces of the baryonic stratum. In essence, both strata are forced into an unstable and temporary state of symmetry that is broken when an up quark is transformed into a down quark (and a positron is created) via the weak force. This induced symmetry can only happen when there is enough concentration of both dark and baryonic matter to sufficiently contract both strata. Brown dwarfs are unable to contract the strata enough to induce interstratum symmetry, and consequently, they are unable to sustain nuclear reactions.

It is already known that the strength of the weak force approaches that of the electromagnetic force at high energies. Within the cores of stars, the combined gravitational influence of the dark and baryonic matters induces symmetry between the electromagnetic and weak forces. In addition, bistratum gravitational symmetry, then, induces symmetry between these baryonic stratum forces and the electroweak force of the dark matter stratum. This force symmetry arises from a stratum-dependent variation in the contraction of space and dilation of time, operating within the two uppermost strata. In isolation, dark matter does not contract space and dilate time in the same manner as baryonic matter because of the stratum-dependent variation in parameters (for example, the constants c and G). Because the strengths of the gravitational force and the derived forces vary with distance, the dark matter stratum must contract disproportionately, more than the baryonic stratum, for bistratum symmetry between the forces to be achieved. The end result is that a dark matter particle is imbued with a great deal of energy and becomes unstable. Energetic particles have greater momentum and oscillate more rapidly through the strata. It is this which permits the exchange of mass (in the form of antimatter) or energy from the dark matter stratum to the baryonic stratum.

Under normal circumstances, dark matter oscillates only within the two lower substrata, which has only eight dimensions of space-time. It is induced to oscillate into the baryonic stratum (and its twelve-dimensional space-time geometry) upon symmetry breaking between the forces of the two uppermost strata. This permits the momentum (which is conserved) of our energetic dark matter particle to be redistributed from two strata to the three in which baryonic matter oscillates. The combined action of the dark matter's electroweak force and the electromagnetic and weak forces of the baryonic stratum forces makes this possible and induces re-expansion of both strata as well. Bi-stratum force symmetry ceases as parametric asymmetry operating within the two strata induces this disproportionate re-expansion of both strata. It is in this manner that nuclear reactions produce positrons (as well as other antiparticles regardless how the bi-stratum force symmetry is achieved, whether in stellar processes or in collider experiments), which are created and permitted to annihilate with electrons in the solar plasma.

As was mentioned previously, the Coulomb barrier is an intrastratum restriction, and it is induced interstratum force symmetry that makes the proton-proton chain reaction possible. Nuclear reactions are a means by which interstratum interaction produces the high-quality energy that is made available for intrastratum gradient reduction. This is accomplished, as one of the protons in the proton-proton chain reaction is converted to a neutron, with the simultaneous production of a positron and electron neutrino (Kutner and Caroll). The W+ particle is the elementary particle that mediates the weak force in nuclear fusion reactions. This particle is produced through induced bistratum symmetry, and it oscillates through the tristratum structure in a different manner than does the photon. Its oscillation through the tristratum structure takes a more complicated path. Unlike the photon, it does not simply oscillate back and forth from the baryonic stratum down through to the dark energy stratum (and repeating the process to form one interstratum wavelength). Instead, during interstratum oscillation, the W+ particle initially materializes within the dark matter stratum when interstratum symmetry is induced. This occurs as one of the up quarks oscillating through the tristratum structure transforms into a down quark during materialization within the dark matter stratum. During this metamorphosis, a W+ particle materializes within the dark matter stratum. The W+ particle then oscillates down through to the dark energy stratum (where it then materializes) and then successively materializes within the dark matter and baryonic matter strata. It then oscillates in the reverse direction through the three strata, and after materializing in the dark energy stratum, it then materializes in the dark matter stratum again. After materializing in the dark matter stratum, it materializes in the dark energy stratum and then successively oscillates (and materializes) through the entire tristratum structure for the second time.

This complicated zigzag course through the tristratum structure represents one wavelength. The final materialization of the W+ particle into the baryonic stratum occurs as bistratum symmetry of the derived forces is broken. (It may be that the induced symmetry between the gravitational forces of the uppermost strata induces symmetry between the Anti-G forces of the two strata as well. Moreover, it may be that the combined bistratum interaction of these oppositional G forces, as well as the complicated route of oscillation just hypothesized, explains the large mass of the W particle.) A positron and electron neutrino are created as the unstable W+ particle rapidly decays as a further consequence of this bistratum symmetry breaking. The newly created positron then annihilates with an electron, producing two gamma ray photons. These two gamma rays carry away energy, which is a bistratum product of interstratum interaction (with an equal contribution from both strata), and this is instrumental in reducing the interstratum gravitational gradient. (The more massive Z particle, which has a neutral electric charge, has a tristratum oscillatory motion, which is a combination of the partial tristratum oscillatory motion of both the W+ and W−, which has not been discussed but differs from that of the W+.)

The second step of the proton-proton chain process is initiated as interstratum symmetry is reinitiated, permitting another proton to fuse with the deuterium produced in the first step. The product of this reaction is a light helium isotope with two protons and one neutron, as well as a gamma ray. The gamma ray produced in this reaction is also a bistratum product of interstratum interaction. In the last step, after millions of years (the first step alone takes approximately a billion years), two light helium isotope nuclei fuse together to form the helium isotope, with which we are most familiar with, which contains two protons and two neutrons. In addition, two free protons are released during this reaction (as well as 12.86 MeV of energy), which are then available for further reactions.

The other two types of proton-proton chain reactions, PP (proton-proton) II and PP III (there is also a PP IV reaction), are less common within stars with mass similar to that of our sun. Nevertheless, these reactions also require the same bistratum interaction previously discussed. The only difference is that these reactions involve higher energies, which requires increased bistratum interaction of the G forces and stratum-dependent derived forces. There will also be an increased contraction (over that involved in PP I reactions) of the bistratum space-time geometry. For stars more massive than the sun (approximately 1½ times the solar mass and higher), the carbon, nitrogen, and oxygen (CNO) cycle becomes the dominant fusion reaction that converts hydrogen into helium. This process involves the utilization of carbon, nitrogen, and oxygen as catalysts in the production of helium.

Moreover, in these more massive stars, bistratum interaction is even stronger than that which takes place in the PP II and PP III reactions. Bistratum contraction (as previously mentioned, there is a disproportionate contraction of the space-time geometry between the two strata, which produces symmetry in the forces involved) is correspondingly greater, which results in nuclear reactions that are significantly more rapid than those involved in the proton-proton chain (the first step alone, in the PP I reaction takes approximately a billion years). This explains why massive stars are relatively short-lived, lasting only millions of years as opposed to the billions of years that represents the lifetime of stars like our sun. The eventual mass of a star will depend on the initial conditions involving bi-stratum gravitational interaction between local concentrations of dark and baryonic matter. In more massive stars, local conditions permit both dark matter and baryonic matter to accumulate in greater quantities before fusion reactions are initiated (further accumulation is then limited in the manner described earlier after fusion interactions are initiated).

In the first step of the CNO cycle (there are also CNO II and OF cycles, which will not be dealt with in this work), induced bistratum symmetry permits a proton to receive the necessary energy to penetrate the Coulomb barrier and fuse with a carbon 12 nuclei to form nitrogen 13. A gamma ray photon is also released during this reaction when bistratum symmetry is broken. In step two, the unstable nitrogen 13 nuclei undergoes beta decay as one of its protons is converted into a neutron producing carbon 13 (see earlier discussion concerning the W+ particle). As was the case in the first step of the PP I chain, a positron and electron neutrino are also released but at a significantly more rapid rate. Once again, the positron released produced as dark matter in the form of antimatter is caused to oscillate into the baryonic stratum, and its twelve-dimensional space-time structure. In addition, two gamma ray photons are produced soon after, as this positron annihilates with an electron within the stellar plasma.

In the third step of the CNO cycle, the carbon 13 isotope fuses with a proton, producing nitrogen 14 and a gamma ray. The nitrogen 14 isotope then fuses with another proton, producing an oxygen 15 isotope, and another gamma ray is also released. The oxygen 15 isotope undergoes beta decay in step five, and one of its protons is converted into a neutron, and as was the case in step two, a positron and electron neutrino are produced as well via the weak force (upon symmetry breaking between the dark and baryonic matter strata). The culmination of the CNO cycle results in the emission of an alpha particle as the nitrogen 15 isotope fuses with a proton and decays to a carbon 12 isotope (Kutner). As previously mentioned, the nuclear reactions of the CNO cycle are significantly more rapid than the PP reactions that take place within less massive stars. Unfortunately, the relative rapidity of this process also partially explains why massive stars use up there fuel so quickly. Nevertheless, nuclear reactions have an important role in hydrostatic equilibrium for all stars and is an efficient means by which bistratum gradient reduction is achieved. The rate and the efficiency of gradient reduction differs between massive and less massive stars, and massive stars do not survive long enough to develop complex planetary systems (suitable for the development of complex life forms). However, they are responsible for producing the rich array of elements that enrich the molecular clouds within the interstellar medium from within which new stars (and planetary systems) are born.

Another type of nuclear reaction, the triple alpha process (also called the Salpeter process), occurs in older stars (red giants) after the proton-proton chain and the CNO cycle have fused most of the available hydrogen into helium. This process requires temperatures of approximately 100 million K (all of the reactions previously discussed are also temperature dependent) and produces carbon 12 through the fusion of three helium nuclei. The high temperature involved in this reaction arises as bistratum interaction of the G forces induces contraction of the dark matter and baryonic strata within a star's core. In the first step, two alpha particles (two helium nuclei) fuse to produce beryllium 8, which is highly unstable and subject to rapid decay (back into two alpha particles). The initial fusion of these alpha particles is initiated as the combined bistratum G forces provide the energy necessary to penetrate the Coulomb barrier. The high temperature within the core, as measured within the baryonic stratum, is also a consequence of the contraction of the baryonic stratum due to the combined influence of the bistratum G forces (recall that the temperature of dark matter cannot be measured unless it is caused to oscillate into the baryonic stratum in the form of antimatter or neutrinos). This permits both alpha particles to be temporarily brought within the range of the strong force operating within the baryonic stratum. Nevertheless, the strong force is incapable of binding the two alpha particles comprising the beryllium 8 nuclei for very long. This may occur because bistratum contraction quickly induces bistratum symmetry between the dark matter's electroweak forces and the baryonic stratum's electromagnetic and weak forces (at the location of the reaction). The combined action of these forces is too strong and quickly overpowers the strong force (as well as the combined bistratum gravitational forces, which induced contraction of the space-time geometry), binding the two alpha particles.

However, production of beryllium 8 is rapid enough that there is always a small amount available, which is able to fuse with an alpha particle to produce an excited carbon 12 nuclei. It may be that in the second step of the triple alpha process, bistratum gravitational interaction is sufficient enough to induce brief symmetry between the strong forces operating within the dark and baryonic matter strata. The combined strength of these forces is more than sufficient to counter the combined strength of the dark matter's electroweak forces and the baryonic stratum's electromagnetic forces. However, the combined interaction of these latter forces is sufficient to expand the bistratum space-time geometry, enough to break the symmetry between the two strata, and end the influence of the dark stratum's strong and electroweak forces. This bistratum symmetry breaking also arises as a consequence of the stratum-dependent variation in parameters, which makes bistratum interaction unstable (as mentioned earlier). Along with the production of carbon 12, gamma ray photons are also produced, and the release of these assists in reducing the *gravitational gradient* at the location of each triple alpha reaction. Hydrostatic equilibrium is maintained so long as a star has fuel to burn to counteract the influence of the bistratum G forces.

Eventually stars use up their fuel and become unstable, and in smaller stars, this instability is alleviated through the ejection of these stars' outer layers in the form of planetary nebulae. The remaining core is prevented from further collapse because of electron degeneracy pressure, and this is how white dwarfs counteract the combined action of the bistratum G forces. The balance between the bistratum G forces and the electron degeneracy pressure effectively prevents sufficient bistratum contraction to induce conversion of the remaining protons in the core into neutrons. In more massive stars, stellar instability is partially alleviated through the expulsion of the outer layers via supernova explosion, with the remaining core collapsing into a neutron star or black hole. The lives of these stars end after the fusion process has culminated in the production of an iron core. This iron core can no longer exert enough outer pressure to support its outer core structure. Electron degeneracy pressure is no longer sufficient to counterbalance the influence of bistratum G forces. The combined action of the bistratum G forces becomes sufficiently strong to initiate an increase in the production of gamma rays without further fusion reactions within the iron core. In addition, the iron core is reduced to its elemental protons and neutrons through photodisintegration as the high-energy gamma rays interact with the iron nuclei (the electrons are already present as ions).

According to current theory, the increased density attained within the core is sufficient to overcome electron degeneracy pressure. Electrons are then *forced* to combine with protons within the core to form neutrons, accompanied by a release of an enormous amount of neutrinos. However, another explanation is possible that involves intensified bistratum interaction, which arises from the inability of the iron core to produce further fusion reactions. In this model, the bistratum gravitational gradient is sufficiently strong to induce conversion of protons within the core to the neutrons without the necessity of combining them with electrons outside the core. Normal fusion processes (as explained earlier) permit a continuous process of local bistratum gradient reduction (at the location of each fusion reaction), which is instrumental in the maintenance of stellar hydrostatic equilibrium. With the cessation of fusion reactions and the photodisintegration of the iron nuclei, this is no longer possible.

Instead, bistratum G forces cause a *global* contraction of the dark matter and baryonic stratum within the core. This results in the simultaneous inducement of symmetry between the dark stratum's electroweak field and baryonic stratum's electromagnetic and weak fields throughout the core. The ensuing instability is alleviated (and bistratum symmetry is broken) once all of the elemental protons are converted to neutrons along with the release of an enormous amount of positrons and neutrinos (an up quark within each proton is converted into a down quark in the manner previously discussed). The release of these neutrinos and positrons, which subsequently annihilate with electrons (producing an enormous amount of gamma rays), relieves the bistratum gravitational pressure sufficiently enough for neutron degeneracy pressure to prevent further bistratum contraction (establishing equilibrium between bistratum G forces and neutron degeneracy pressure upon core rebound). When the ensuing core rebound takes place (due to the bistratum release of mass and energy), the resulting shock wave, along with the outwardly flowing high energy gamma rays and neutrinos, reverses the direction of the inwardly falling matter of the outer layers.

During this supernova explosion, the expelled outer layers of the star are compressed by the shock wave and the release of gamma rays and neutrinos. This leads to the rapid contraction of both the dark matter and baryonic strata in the location of the outer layers. Moreover, this induced contraction results in a rapid process of nucleosynthesis (and bistratum re-expansion as bistratum field symmetry is broken) and the production of the elements with atomic numbers (and masses) greater than iron. The rapid production of neutrons during this process results in a further production of gamma rays (through positron-electron annihilation) and neutrinos. The combined contribution of gamma rays produced via core collapse and outer layer nucleosynthesis accounts for the reason that a supernova can briefly light up the night sky brighter than a galaxy.

The remnant core that comprises the newly formed neutron star has reached what is called its Oppenheimer-Volkov limit, which establishes the upper limit of neutron star mass (Comins and Kaufmann). Above this limit, neutron degeneracy pressure will not be sufficient to counterbalance bistratum G forces and prevent further core collapse. It is theorized that a type of star with a mass that exceeds the Oppenheimer-Volkov limit may exist, which is called a quark star. Although quarks are normally confined within neutrons (as well as protons and mesons) via the strong nuclear force, within a quark star gravitational pressure is sufficient to decompose the star's neutrons into individual quarks. Such a star, if it exists (which is still speculative), is prevented from further gravitational collapse because of quark degeneracy pressure, in accordance with the Pauli exclusion principle (Comins). As was the case with neutron stars, these quark stars are the remnant product of supernovae explosions. One difference between the supernovae explosions of neutron stars and that of quark stars may be in the total amount of energy released when they occur. Another difference may be in the extent to which they contract the bistratum space-time structure, which still falls short of complete collapse.

However, if the collapsed stellar remnant of a massive star possesses enough mass quark degeneracy pressure (or any other theoretical degeneracy pressure) will not be sufficient to counterbalance bistratum gravitational pressure. The result is the creation of a black hole, which is probably the least understood of all celestial objects, and most of what is believed about them is probably wrong. This is, in part, because of the incompleteness of general relativity and the absence of a quantum theory of gravity (and a corresponding theory about the quantization of space-time, all of which was discussed earlier), as well as ignorance of the existence of the short-range Anti-G force and the stratification of the space-time geometry. Another reason for our current lack of understanding is because they are difficult to study because their presence can only be inferred by the effects they have on surrounding material. In the previous discussion about the various types of stars, including those in hydrostatic equilibrium, as well as remnant stars like white dwarfs, neutron stars, and quark stars, there exists in each case a means by which complete tristratum collapse is prevented. This is accomplished as bistratum gravitational forces are halted by opposing forces (the derived forces), limiting the contraction of the tristratum space-time geometry. Moreover, this permit's the particles that comprise these celestial objects to continue oscillating into their respective strata. This means that dark matter particles are permitted to continue oscillating within the two lowermost substrata and its combined eight-dimensional structure. At the same time, particles inhabiting the baryonic stratum continue to oscillate through the entire tristratum (twelve-dimensional) structure.

The situation is entirely different for black holes, as the stratum-specific derived forces are overpowered by the multistratum G forces. What results is a complete collapse (as mentioned earlier) of the tristratum structure within the region of space in which the black hole forms. This means that all of the particles trapped within these regions during collapse cease to continue to oscillate within their stratum-specific space-time structure. All interstratum and intrastratum motion has ceased, and the entire black hole structure has achieved what will be henceforth called *Perfect Zero temperature*. As mentioned earlier, only intrastratum measurements of temperature can be experimentally determined. This is because our measuring devices are classical objects that comprise an immense number of particles oscillating within the entire twelve-dimensional space-time geometry. Measurements of the temperatures of neither dark matter nor dark energy can be made directly because neither type of matter normally oscillates into the baryonic stratum (this does not discount indirect measurements of their effects on baryonic matter as a consequence of interstratum interaction). This also means that there can be no measurement of the intrastratum accumulation of entropy (a topic that will be addressed shortly) within the lower substrata either. In other words, from the perspective of a classical observer (a scientist perhaps), both dark matter and dark energy have no temperature and not simply *absolute zero* under normal circumstances (that is, unless it is caused to oscillate into the baryonic stratum, as occurs in the fusion reactions discussed earlier).

According to current theory, *absolute zero* is the temperature at which matter has what is called zero-point energy, and it is also the temperature at which the entropy of a system has its minimum value. At absolute zero, particles possess a minimum of kinetic energy but are not capable of transferring any kinetic energy to other particles (Hewitt). In this sense, the concept of absolute zero is classical because it is a measure of the activity of particles that oscillate into the baryonic stratum (it is an intrastratum measurement). This implies that particles continue to have interstratum oscillatory motion, and although this motion is at a minimum, it has not ceased altogether. In contrast, there is no interstratum oscillatory motion taking place within black holes (because gravistatic equilibrium makes it impossible) and hence the need for the adoption of a new concept of minimal temperature, which prevents confusion in terminology. At the temperature of *Perfect Zero*, all interstratum and intrastratum motion has ceased (time stops), and intra-stratum entropy is minimal within each stratum as well. This means that black holes have minimal entropy and, contrary to current theory, are the lowest-entropy objects in the universe. It also means that during the expansion phase of the universe, black holes can never decrease in size (sorry Stephen).

Hawking's ideas concerning black hole evaporation are based on quantum field theory. According to this theory, virtual particles are continuously "popping in and out of existence" (Carroll) within the vacuum. It is believed that virtual particle-antiparticle pairs spontaneously appear, and disappear (through annihilation) rapidly and that the pair possesses a net energy of zero. Hawking concluded that these *virtual particles* can be transformed into real particles in the presence of the gravitational field of a black hole. For the net energy of these particles to remain zero, it is necessary that one of the pair must possess a positive energy and the other must possess negative energy. It is further contended that the newly created particle with negative energy is captured by the gravitational field and penetrates within the event horizon (decreasing the mass of the black hole). In the meantime, the particle with positive energy escapes, and this is the manner in which it is alleged that black holes evaporate via Hawking radiation (Carroll).

The problem, as I see it, comes from our lack of a complete understanding of quantum processes, as well as our lack of understanding of the actual cause of the uncertainty, which the uncertainty principle attempts to describe. (For example, the uncertainty principle does not apply within black holes because all oscillatory motion ceases in the region inside the event horizon.) This work has already suggested that it is the heretofore ignorance of the complexity of space-time structure that has led to the present state of confusion, as well as the theoretical flights of fancy, which have been embraced all too readily by the orthodox community. Of course, this author concedes that she may be just as guilty as those who she disagrees with, but it is the truth that she seeks regardless of where it may lead. It is increasingly clear that the current state of affairs requires revolutionary thinking and the type of risk-taking not afforded to those who desire professional security and group acceptance. However, with the knowledge that today's heresy may become tomorrow's orthodox *truth*, this author will proceed boldly onward.

With this in mind, it is time to step down from the soapbox and return to our discussion on black holes and why they *do not evaporate nor possess high entropy* (or for that matter, electric charge—I know I should be burned, wretched heretic that I am). Early on, it was hypothesized that what we normally recognize as the vacuum can be identified with the dark energy stratum, and its four-dimensional space-time structure. Because of its variation in stratum-dependent parameters, the current repulsive property of its combined strong-electroweak force, and its limited four-dimensional structure, it only weakly influences the activity of both dark matter and baryonic matter. It does, however, provide an increasingly expanding environment (the cause of this expansion will be explained later) in which the complex structures that comprised these types of matter (dark matter and baryonic matter) become increasingly localized and isolated as the universe ages. It is within the vacuum that quantum field theory postulates that *virtual particles* are continuously "popping in and out of existence" (Carroll), making what is normally regarded as empty space an environment in which there is constant activity.

The problem with quantum field theory, as it is currently understood, is that it does not incorporate the complexity of the multidimensional structure within its framework. For example, the electromagnetic force is mediated via exchange of virtual photons, which are oscillating through the baryonic stratum's entire twelve-dimensional structure. Moreover, as mentioned earlier, these virtual photons have stratum-dependent variations in intrastratum velocity during interstratum oscillation (as they materialize within each stratum), which explains how they can be exchanged at velocities greater than c, as measured in the baryonic stratum. Normally, these virtual photons cannot be detected because during a significant amount of their travel between charged particles, they are traveling faster than the speed of light c, as it is measured within the baryonic stratum.

Both electrons and protons permeate space at varying distances from each other; therefore, space is continuously flooded with virtual particles in all directions, and these comprise what has been traditionally called the electromagnetic field. In regions of space where there is a high concentration of charged particles, the electromagnetic field is strong. This is because there are more virtual photons exchanged with a wide variation of interstratum oscillatory momentum, with those exchanged at the shortest distances, having the greatest interstratum momentum. In regions of lower concentration of charged particles, the increased distances between particles result in an exchange of particles with lower interstratum momentum. Moreover, the interstratum momentum of the virtual photons is inversely proportional to the distance of the exchange between individual pairs of charged particles (as you recall, the minimum possible value for interstratum momentum places a limit on the range of the electromagnetic force). This is why the strength of the electromagnetic field is weaker in otherwise empty space because only virtual particles with low interstratum momentum are present. [Dark matter also exchanges virtual particles, which constitute the dark matter stratum's electroweak field. These exchange particles also vary in intrastratum velocity as they oscillate through the dark matter and dark energy strata. In contrast, dark energy particles only oscillate within their own stratum, which means that there is seldom a variation in interstratum or intrastratum momentum except under interstratum gravitational influence (and no virtual particle exchange in the sense described for the other two types of matter). This explains why the density of dark energy remains relatively unchanged, except under circumstances where concentrations of dark matter and baryonic matter introduce sufficient gravitational influence to counteract the strong-electroweak force of the affected dark energy.]

The variation in intrastratum momentum that virtual photons undergo results in another significant difference in their motion from that of real photons in the presence of strong gravitational fields. As mentioned earlier, real photons have no variation in intrastratum velocity as they travel through space, which means that they are more susceptible to changes in trajectory in the presence of a strong gravitational field. In contrast, virtual photons will almost always travel in a virtually straight path, except in the presence of a tremendous gravitational field like that in the vicinity of a black hole. This is because during most of their travel, virtual photons are traveling faster than the velocity c as measured within the baryonic stratum. However, in the vicinity very close to a black hole, the space-time geometry of the tristratum structure is contracted to varying degrees. This results in a slowing down of the virtual photon during its intrastratum velocity when materializing within the lower strata. The intrastratum velocity of the virtual photon then becomes that of the velocity c as measured within the baryonic stratum, when traveling through all three strata. Consequently, the virtual photon ceases to be virtual, and the newly created real photon will have an energy corresponding to the interstratum momentum it possessed before its metamorphosis.

It is in this manner that charged particles release some of their energy and lose angular momentum before being swallowed up by a black hole. The real photons created in this process with the greatest energy (and higher frequencies) are those that are transformed during exchange between charged particles at increasingly shorter distances. This results in a collapse of the electromagnetic field (as well as the electroweak and strong-electroweak fields) for those particles that penetrate the event horizon as they shed energy and momentum. Pair production of real positrons and electrons may occur if newly created real photons (metamorphosed virtual photons) have sufficient energy (that have enough interstratum momentum). However, since the interior of a black hole is in gravistatic equilibrium and the derived fields of the tristratum structure have collapsed, there is no net electric charge within the black hole interior (and no magnetic properties either). This means that for electric charge to be conserved, it is necessary that pair production must be followed by immediate annihilation before penetration of the event horizon transpires. It also means that a normal matter of opposite charge (protons and electrons) must penetrate the event horizon in equal amounts as well.

There is no such thing as negative energy (or mass), and it there is limited knowledge about quantum behavior, which has given rise to this idea in the first place. Real particle creation through the metamorphosis of virtual particles is always the result of the loss of energy from interacting charged particles, which takes place outside the event horizon. Furthermore, this energy is always positive, and there is no means by which a black hole may lose mass (nor can gravistatic equilibrium be disrupted) during the expansion phase of the cyclic process. Evaporation via quantum tunneling is not possible either because the complete collapse of the tristratum structure within black holes into a state of gravistatic equilibrium makes it impossible for this phenomenon to occur. Penetration of the Coulomb barrier (described earlier) in stellar processes involves the complex interaction between the gravitational and derived forces of the dark and baryonic matter strata. Within black holes, these forces (along with the antigravitational force) are in a perfect balance, which prevents all oscillatory behavior from taking place (there are no longer identifiable charged particles, just collapsed fields within collapsed units of quantized space-time) within the black hole (which means they cannot radiate—the electromagnetic field is no longer operating).

Another important point to consider is the notion of alleged black hole entropy. Normally, for most systems, entropy is proportional to the volume of the given system (a significant fact you might think). However, Hawking (and Beckenstein) concluded that the entropy of a black hole is proportional to the area of the black hole's event horizon (Greene). It may very well be that Hawking has derived a correct formula for the entropy at or near the event horizon. However, this does not mean that it applies to the region of space within. If the event horizon is just a barrier between the collapsed space-time geometry within and the (uncollapsed) outside universe, then Hawking's formula may be valid. Of course, that is only true if the event horizon itself represents a barrier of space-time that is not fully collapsed (into gravistatic equilibrium). In other words, if oscillatory motion does not cease at the event horizon (proper), then the event horizon will have a measurable amount of entropy consistent with Hawking's calculations. In contrast, the black hole interior will have a structure (constituting the volume, extending to the event horizon) consisting of an incredible number of quantized units of fundamental space-time at *Perfect Zero* temperature (as described earlier) and in a state of gravistatic equilibrium (with the lowest possible entropy). (Black holes may have a layered, onionlike structure, with the collapsed remains of each of the three strata existing in separate layers composed of the quantized units of space-time discussed earlier in this work.)

The cascading collapse of the tristratum structure of space-time (within black holes) is the process by which interstratum interaction eliminates intrastratum entropy accumulation. (The interstratum collapse of the hierarchically stratified space-time structure effectively erases any accumulated intrastratum entropy. This includes those during initial black hole collapse and all materials that subsequently penetrate the event horizon.) The interstratum interaction between the dark and baryonic strata in stellar processes is an efficient form of work (which is interstratum in character), which produces the high-quality energy available for intrastratum gradient reduction. As mentioned earlier, real photons may only interact with particles (which also oscillate into the baryonic stratum) during materialization within the baryonic stratum (only virtual photons can interact during materialization when oscillating within the lower strata). Therefore, the energy of a photon is its energy as it is measured during oscillation into the baryonic stratum.

In accordance with the First Law of thermodynamics, energy can be transformed into other forms. The electromagnetic energy from the sun is stored in plants via photosynthesis and later converted to the chemical energy contained in fossil fuels (after millions of years). Regardless of the form energy may take, any measurement of its *exergy content* is an intrastratum measurement. In addition, intrastratum gradient (thermal) reduction involves work performed at the intrastratum level between organizations of particles during oscillation within the baryonic stratum. As was mentioned earlier, *temperature is an intrastratum measurement*, which can only be made on particles during oscillation within the baryonic stratum. A heat engine is an example of a macroscopic object, which is composed of an enormous amount of particles that are constantly oscillating out of step through the tristratum structure. Heat engines produce work through reduction of a temperature gradient between a *hot source* and what is referred to as a *cold sink*. Moreover, both the temperatures of the hot source and cold sink are intrastratum properties, as is the entropy produced as work is performed. (Irreversible processes are generally intrastratum phenomena; one exception is the interstratum production of positrons and gamma rays in stars in the manner described earlier. On the other hand, reversible processes involve interstratum interaction or activity.)

Hence, when one says that the entropy of a system is proportional to its volume, what is being referred to are the microstates of the particles contained within as they oscillate into the baryonic stratum (and not their behavior during lower-stratum oscillation). Inside a black hole, there is a complete absence of intrastratum activity because there is no longer interstratum oscillation in any of the three strata (as a consequence of gravistatic equilibrium) and the 2nd Law of thermodynamics no longer applies. Gravistatic equilibrium only preserves the most basic information, information concerning the parameters associated with each strata. In essence, black holes (which are in a *dormant* state, much like a seed) carry within them the *cosmogenetic* information, that serves as the blueprint for the fundamental developmental structure of the next expansion cycle.

Interstratum collapse of each of the three strata effectively erases the intrastratum entropy, that has been accumulated during the cosmic expansion phase.

Another method by which some of the intrastratum entropy of the baryonic stratum may be reduced is via a form of oscillatory decay. This may occur as photons (and perhaps neutrinos) that normally oscillate through the tristratum structure reach a minimum threshold of energy. There may be a maximum or longest wavelength, which a photon may possess, that establishes the minimal limit of interstratum momentum a photon requires to continue to oscillate through the tristratum structure. Photon energy is continuously reduced as energy is used to perform work within the macroscopic structures that *inhabit* the baryonic stratum. As was mentioned earlier, the gravitational influence of dark energy may also play a role in diminishing photon energy as well (because photons travel slower than the constant c, as measured within the dark energy stratum). As a consequence, energy that is rendered useless for activity within the baryonic stratum may be transformed and made suitable for use within the dark energy stratum.

In other words, the former photon, which had been oscillating into the twelve-dimensional tristratum structure associated with the baryonic stratum, is transformed into dark energy (in accordance with a new law of thermodynamics, the *Law of Entropy Reduction via Interstratum Conversion*). In its new form, the newly created dark energy particle will proceed to only oscillate within the four-dimensional structure of space-time associated with the dark energy stratum. In this manner, a portion of the entropy of the baryonic stratum is adapted for use in increasing cosmic expansion. Since this is a continuous and cosmicwide process, the amount of dark energy in the universe is steadily increasing. This also explains how the density of dark energy may remain constant, even though the expansion of space is accelerating. In fact, the currently observed acceleration may be, in part, a consequence of an escalating increase in photon to dark energy conversion (and perhaps neutrino to dark energy conversion) as the universe ages.

The primary function of the dark energy stratum is to provide an environment in which interaction between the dark matter and baryonic matter strata can produce structures, which reduce the cosmic gradient. As an expanding environment, it fosters the localization of structure development, which permits the emergence of variation and complexity of organization. As mentioned previously, it also serves as the *cold reservoir* (or sink) into which *waste heat* produced through the thermodynamic processes involved in interstratum interaction between the dark and baryonic strata is discarded. Through expansion, it also limits the local accumulation and density of both dark matter and baryonic matter. This, in turn, prevents the thermal gradient from becoming too steep in a given region of space. It also slows down the process of black hole formation, which furthers the potential for the development of complex structure. Because the interaction of the dark and baryonic matter strata produces the photons, which are eventually converted to dark energy, the rate of cosmic expansion is interdependently determined throughout the cosmic lifecycle (more about that shortly). It is the nature of this interdependent relationship that changes as the universe evolves during the course of cyclic evolution. Parametric mutagenesis and the mitotic or fissionlike division of the space-time geometry into *separate strata* are responsible for the changes in this interdependent relationship.

Another factor that may contribute to the maintenance of constant dark energy density is the nature of the strong-electroweak force, which operates within the dark energy stratum. It may well be that this force is a short-range force like the strong and weak forces that operate within the baryonic matter stratum. If so, like the strong force, the strong-electroweak force may be strongest at its maximum range (and repulsive) and become steadily weaker at shorter distances. There may be a minimum distance at which this force ceases to be repulsive altogether and instead becomes an attractive force (a switch in monopolarity at a threshold distance), which increases in strength at ever shorter distances. In most cases, the repulsive character of this force predominates, and this helps maintain constant density of dark energy throughout most of the universe. Even in regions of space occupied by massive macroscopic bodies like planets and stars, the combined gravitational forces of the local concentration of dark and baryonic matter is insufficient to overcome the repulsive form of the strong-electroweak force.

However, in regions of space where black holes collapse the entire stratified structure, the dark energy stratum within this region is contracted to a short-enough distance that the strong-electroweak force becomes attractive in nature. This then facilitates collapse of the dark energy stratum and its constituent dark energy particles, which happen to be contained within the black hole's event horizon during the process of gravitational collapse of the tristratum structure. Even after black hole collapse, any dark energy particles that approach the event horizon may reverse monopolarity if they approach close enough. In this manner, dark energy particles continue to be slowly (as is any dark matter that approaches close enough) consumed by black holes. However, because new dark energy particles are being created constantly and in greater number than those consumed by black holes, space continues to expand.

It is now believed that approximately 5 billion years ago, space expansion began to accelerate. There may be several factors involved in bringing this about. Five billion years ago, concentrations of dark and baryonic matter were closer to each other and were capable of exerting greater gravitational influence between each other, as well as the surrounding dark energy. Because of this stronger gravitational interaction, the density of dark energy was probably greater than it is today. This is because interstratum gravitational influence was sufficient to contract the dark energy particles and weaken the repulsive force between them (but not sufficient to reverse monopolarity). Nevertheless, the increase in dark energy production as photon (and perhaps neutrino) decay steadily increased proceeded to slowly cause greater separation between localized concentrations of dark and baryonic matter (primarily associated with galaxies). The weakening of intergalactic gravitational influence (with increasing separation) permitted the strong-electroweak force to operate at increasingly greater strength and expand the dark energy in the intergalactic vacuum. In addition, there was a continual loss of both dark matter (through photon and neutrino creation) and baryonic matter through photon (and perhaps neutrino) decay. This meant that there was increasingly less of both types of matter to exert gravitational influence on the ever-increasing supply of dark energy.

Of course, if this author is correct, there may be a limit to the amount of dark energy that may be produced during the current expansion cycle. If the source of increase is due to photon (and possibly neutrino) decay, then when all baryonic matter and dark matter have been consumed in black hole formation (that is when all of the fields of the dark and baryonic strata throughout infinite space have collapsed into a state of gravistatic equilibrium), no more dark energy may be produced. This means that cosmic expansion must halt at some point because of the cessation of dark energy production and, as a consequence of the previously described, the limited range (which is short) of the strong-electroweak force. Once cosmic expansion has halted; the process of gravitational collapse of the dark energy stratum will follow as the dark energy particles throughout infinite space reverse monopolarity.

This may begin relatively slow at first, as the space in the regions nearest the widely dispersed black holes begins to collapse with local reversal of monopolarity. The dark energy's strong-electroweak field will then become attractive in nature (recall that the value for the constant c is highest in the dark energy stratum, and this is the velocity at which the strong-electroweak force and the dark energy's G force is transmitted), and this will expedite local collapse of the dark energy stratum. What will follow will be an increasingly rapid chain reaction-like escalation of monopolarity reversal throughout the remainder of the dark energy stratum (and its four-dimensional space-time structure). This will mark the end of the current expansion phase, as cosmic collapse is followed by a rapid initiation of cosmic re-expansion (the next big bang). This occurs because the collapsing dark energy stratum (except that which was trapped during black hole formation) never achieves a state of gravistatic equilibrium, and instead, the unstable relationship between the G and Anti-G forces is permitted to manifest itself almost immediately.

Instability between these oppositional G forces is brought about through the combined attractive nature of the strong-electroweak force (which had reversed its monopolarity) and the G force operating in the dark energy stratum. Tremendous pressure rapidly builds up, and the temperature escalates until it reaches the Planck temperature associated with the dark energy stratum. The Planck temperature is higher for the dark energy stratum than it is for the other two strata because of the higher value for the constant c and the lower values for the constant G and the reduced Planck constant. In addition, the Planck time is shortest for this stratum as well for the same reason. This means that the dark energy stratum will expand first at the onset of the next cosmic cycle once this stratum has reached its Planck temperature, and it will occur at this stratum's Planck time. (As mentioned earlier in this work, there are no singularities; it is infinite space that expands and contracts.)

Cosmic collapse of the dark energy stratum introduces instability into the black holes, which contain all three strata in a state of gravistatic equilibrium. Gravistatic equilibrium is disrupted in the dark energy stratum within the black holes first. This will follow after the initial expansion of the largest portion of the dark energy strata that had never attained gravistatic equilibrium during the cosmic collapse process. At this early time, expansion is rapid and resembles the inflationary period predicted in current models because both the dark matter and baryonic strata still remain (however brief) in gravistatic equilibrium and are not yet gravitationally active. There is also some interstratum exchange of energy (which is instrumental in disturbing the state of gravistatic equilibrium in the upper two strata and in inducing re-expansion) from the dark energy stratum during re-expansion of the dark and baryonic matter strata. In this manner, the dark energy stratum returns the energy it received through the process of photon (and perhaps neutrino) decay in the previous cycle. As the rapidly expanding dark energy stratum cools to the Planck temperature of the dark matter stratum (and its associated Planck time), gravistatic equilibrium is disrupted in this stratum (the dark matter stratum), and it expands. (Gravistatic equilibrium is a state that is only permitted to occur locally during the expansion phase. This is because the gravitational influence of the material between black holes prevents the local curvature of space within the event horizon from becoming sufficient to induce unstable interaction between the G and Anti-G forces. For example, the gravitational attraction of the material (both baryonic and dark matter) that comprises our galaxy exerts a gravitational influence on the supermassive black hole at its center. All of the material outside the galaxy, including distant galaxies and the vast expanse of intergalactic dark energy, also exert a gravitational influence on our galaxy's supermassive black hole. Complete cosmic collapse at the end of the cosmic cycle disrupts this delicate balance.)

As gravistatic equilibrium is terminated in the dark matter stratum, this stratum begins to exert gravitational influence on the expanding dark energy stratum, slowing its rate of expansion. With continued expansion of this stratum, it begins to cool along with the dark energy, and once the combined structure cools to the Planck temperature associated with the baryonic matter stratum, this stratum then expands as well (at its associated Planck time). Once the baryonic stratum is reanimated, it also begins to exert gravitational influence on both the dark energy and dark matter present. This results in a further slowing down of the expanding dark energy stratum and the reintroduction of structure via interaction between the dark and baryonic matter. The reintroduction of the influence of the baryonic stratum's G force promoted the accumulation of the dark matter that had been widely dispersed before that time. The high densities of dark and baryonic matter (and the dark energy due to interstratum gravitational influence) at this early time also resulted in the strong interstratum interactions that produced the early supermassive black holes around which galaxies developed. In the early universe, the relative densities of all three types of matter facilitated interstratum gravitational collapse. Supermassive black hole formation (as well as black holes of smaller size, but not mini-black holes that cannot exist) only began to slow as the universe continued to expand, which lessened interstratum interaction. The resulting reduction in interstratum gravitational influence also permitted the development of smaller structures such as stars. Of course, none of this could occur until after symmetry breaking permitted the separation and differentiation of the forces in each stratum in the early moments of the universe.

The symmetry breaking of the various forces associated with each stratum probably occurs at some intermediate (intermediate between Planck temperatures of successive strata) temperatures and times, which are stratum dependent. (Revitalization of the derivative forces permits the *G force* to operate at a greater distance and with lesser intensity in relationship to the Anti-G force.) For example, the symmetry breaking that resulted in the separation of the strong-electroweak force from the G and Anti-G forces within the dark energy stratum (at the termination of gravistatic equilibrium) probably occurred at a temperature higher than the Planck temperature associated with the dark matter stratum. The symmetry breaking (producing the separation and differentiation of its derivative forces) associated with the baryonic matter stratum did not occur until after symmetry breaking within the lower strata was complete.

Ultimately, it is the stratum-specific parameters associated with the Anti-G and G forces that determine the Planck temperatures and Planck times at which stratum expansion occurs. In addition, it is the stratum-specific parameters (which were preserved during gravistatic equilibrium in the previous cycle) associated with each stratum that determine the temperatures and times at which derivative (or offspring) force separation and differentiation occurs as well. With complete reanimation of the strata, there is a reintroduction of interstratum oscillatory behavior or, in the case of the dark energy stratum, intrastratum oscillatory behavior (within its four-dimensional space-time structure).

The transcyclic transfer of information (*without transcyclic transfer of entropy*) allows for a measure of *transcyclic continuity*, as well as measure of stability in the cyclic process. This cosmic hereditary transfer (of fundamental parameters) ensures that the universe does not need to start from scratch each cycle in response to the inherent instability between its oppositional G forces. However, this process is not perfect and mutations can arise spontaneously when critical thermodynamic thresholds are reached. After all, the universe can never permanently relieve itself of the fundamental instability that drives its development and evolution. The transitory nature and the novelty of the structures it evolves and develops are a consequence of this inherent instability, and it is also the reason why we exist.

In accordance with the Law of Maximum Entropy Production (which should probably be called the Law of Maximum Efficiency of Entropy Production), the interactive and interdependent relationship between the three strata of space-time structure continues to evolve in a manner that increases the efficiency of gradient reduction. This means that there may be slight increases in efficiency from cycle to cycle until a maximum limit is reached, and then the process begins anew, as described earlier in this work. The rate at which the universe expands (and the duration of the expansion phase) before it halts at the termination of the cycle is dependent on the efficient interdependent interaction of the strata. It is this that determines the rate at which entropy produced through interstratum interaction between the dark and baryonic strata is converted to dark energy. The shorter duration of more primitive cycles may be attributable to a more rapid formation of black holes that limited production of dark energy. This meant that with early termination of dark energy production, a halt to cosmic expansion would come quicker and so would cosmic collapse.

It may be that random changes that result in more efficient entropy production and gradient reduction via interstratum interaction between dark and baryonic matter during a cosmic expansion cycle may be responsible for parametric mutagenesis in a subsequent cosmic cycle. Changes in the amount of dark energy produced before cosmic collapse may impact the relationship between the G and Anti-G forces operating within the dark energy stratum upon the collapse and re-expansion. This may result in changes in the Planck temperature (and Planck length) and Planck time at which the dark energy stratum re-expands. It may also lead to an alteration of the parameters associated with the forces operating within this stratum. This, in turn, impacts the amount of energy transferred to the other strata when gravistatic equilibrium is terminated, as well as the temperatures and times at which this occurs. Changes in the strength and the range of the strong-electroweak force may affect the density of dark energy during cosmic expansion. There may also be changes that affect the circumstances under which monopolarity reversal occurs. In more primitive cosmic lifecycles, the reversal from a repulsive to an attractive property of the strong-electroweak force probably occurred with greater rapidity, leading to a shorter cosmic lifespan.

It is probable that changes in parameters, when they do occur, are relatively small but, nevertheless, have a significant impact on the evolution and development of the universe as the cosmic lifecycle proceeds. In addition, because of the interdependent nature of the cosmic structure, thermodynamically driven selection processes will favor those changes that increase the thermodynamic efficiency of the entire structure. The observed fine-tuning of parameters and the low entropy state of the early universe cannot be explained within the context of a cyclic model, which allows for the transcyclic transfer of entropy (thermodynamic refuse). The only cyclic model that is satisfactory is one in which the state of our universe can be explained within the context of a long, cyclic evolutionary history. This requires that the 2nd Law of thermodynamics must have a more restricted role in cosmic evolution.

The Law of Entropy Reduction via Interstratum Conversion provides the mechanism by which the role of the 2nd Law is restricted. This is accomplished through the interstratum process involved in the production of dark energy, as well as the elimination of accumulated intrastratum entropy via the interstratum collapse process involved in black hole formation. The latter process represents the ability of oppositional G force interaction (between the gravitational and antigravitational forces operating within the tristratum structure) to transform waste energy and recycle it for more productive use in the succeeding cosmic lifecycle. The postulated existence of a multiverse with a plethora of parameters does not provide an explanation of how our universe acquired the parameters it has, nor is such an explanation consistent with any reasonable notion of causality. Instead causality is abandoned and replaced with blind probability, which represents the easy way out. One would think that it would be far better for physicists and cosmologists to pursue an *autocausal explanation* that is based on properties of our own universe. Until this takes place, the hypothetical multiverse will represent nothing more than a cosmological cop out of multi-universal proportions.

One of the motivations for writing this work was a growing sense of dissatisfaction with the direction cosmology (as well as quantum theory) appeared to have taken over the last several decades. Many of the ideas put forth by otherwise respectable scientists (who I concede are smarter and more knowledgeable than I am): the multiverse, the proposed existence of singularities, the Big Rip, and the notion of an ultimate beginning at the big bang to name but a few, appeared to make little sense. These ideas and others appeared to lend support to the notion of a universe on LSD or one which does not differ much from that portrayed in Rod Serling's *Twilight Zone*. No doubt, many of these ideas have provided excellent fodder for the imaginations of science fiction authors and screenwriters. However, I do not believe that they have contributed much in the search for understanding. Perhaps, those presented in this work are no less speculative and outlandish than those of others who have had the benefit of extensive training. Nevertheless, a darkness has fallen on the world of science, and it would seem that we may be on the threshold of a revolution in physics and cosmology, as well as a much needed and thorough house cleaning. It would seem almost inevitable that many currently popular ideas discussed in the scientific literature are destined for an ignoble end and to be discarded remorselessly into the dustbin of history.

In any case, it has become abundantly clear that many of the problems confronting cosmology will require a rethinking and a revolutionary makeover that is more than cosmetic. When all is said and done, I think that it will be found that the universe does indeed *make sense*. By that, I mean that she will be found to be infinite in extent, self-perpetuating, eternal, and forever evolving via a cyclic process. It is hoped by this author that the ideas contained in this work may provide a direction and a starting place. In addition, in view of the nation's current economic crisis, this work can be regarded as a possible *economic stimulus* for authors of textbooks and popular science books. Or then again, maybe not.

References

Anastopoulos, Charis. *Particle or Wave*. Princeton University Press, Princeton, 2008.

Atkins, Peter. *Four Laws That Drive the Universe*. Oxford University Press, New York, 2007.

Barrow, John D. *The Constants of Nature from Alpha to Omega*. Pantheon Books, New York, 2002.

Barrow, John D. *New Theories of Everything*. Oxford University Press, New York, 2007.

Bertalanffy, Ludwig von. *General Systems Theory. Foundations, Development, Applications*. George Braziller, Inc., New York, 1969.

Bethe, H.A. *Energy Production in Stars*. Physical Review, vol. 55, 1939.

Bondi, Hermann. *Relativity and Common Sense: A New Approach to Einstein*. Doubleday & Company, Inc., 1964. Dover Publications, Inc., Mineola NY, 1980.

Born, Max. *Einstein's Theory of Relativity*. Dover Publications, Inc., New York, 1965.

Calle, Carlos I. *The Universe—Order Without Design*. Prometheus Books, Amherst NY, 2009.

Chown, Marcus. *The Quantum Zoo A Tourists Guide to the Never-Ending Universe*. Joseph Henry Press, Washington DC, 2006.

Clegg, Brian. *Before the Big Bang The Prehistory of Our Universe*. St. Martins Press, New York, 2009.

Clegg, Brian. *The God Effect Quantum Entanglement, Science's Strangest Phenomenon*. St. Martin's Press, New York, *2006*.

Close, Frank. *Antimatter*. Oxford University Press, New York, 2009.

Comins, Neil F. and Kaufmann, William J. *Discovering the Universe*. W. H. Freeman and Company, 7th edition 2005. 8th edition, New York, 2008.

Cox, Brian and Forshaw, Jeff. *Why Does E = mc squared? (And Why Should We Care?)*. Da Capo Press, Philadelphia, 2009.

Darling, David. Gravity's Arc *the Story of Gravity, from Aristotle to Einstein and Beyond*. John Wiley & Sons, Inc., Hoboken NJ, 2006.

Darwin, Charles. *On the Origin of Species*. Gramercy Books distributed by Random House Value Publishing, Inc., New York, 1979.

Davies, Paul. *The Cosmic Blueprint*. Templeton Foundation Press, Paperback edition, Radnor Pennsylvania, 2004.

Davies, Paul. *Cosmic Jackpot (Why Our Universe is Just Right for Life)*. Houghton Mifflin Company, New York, 2007.

Dosch, Hans Günter. *Beyond the Nanoworld Quarks, Leptons, Gauge Bosons*. A K Peters, Ltd., Wellesley MA, 2008.

Einstein, Albert. *Relativity (Great Minds Series)*. Prometheus Books, Amherst NY, 1995.

Fermi, Enrico. *Thermodynamics*. Prentice-Hall Company, 1937. Dover Publications, Inc., New York, 1956.

Ferris, Timothy. *The Whole Shebang*. Simon & Schuster, New York, 1997.

Feynman, Richard P. *QED the Strange Theory of Light and Matter*. Princeton University Press, Princeton, 2006.

Feynman, Richard P. *Six Easy Pieces*. Basic Books, New York, 1995.

Feynman, Richard P. and Leighton, Robert B. and Sands, Matthew. *The Feynman Lectures on Physics* vols. 1-3. Addison-Wesley Publishing Company, Reading MA, 1965.

Ford, Kenneth W. *The Quantum World-Quantum Physics for Everyone*. Harvard University Press, Cambridge MA, 2005.

Frampton, Paul H. *Did Time Begin? Will Time End?* World Scientific Publishing Co. Pte. Ltd., USA office—Hackensack NJ, 2010.

French, A.P. *Vibrations and Waves*. W W Norton & Company Ltd., New York, 1971.

Freeman, Ken and McNamara, Geoff. *In Search of Dark Matter*. Praxis Publishing Ltd., Chichester UK, 2006.

Fritzsch, Harald. *The Fundamental Constants: A Mystery of Physics*. World Scientific Publishing Co. Pte. Ltd., USA office—Hackensack NJ, 2009.

Gamow, George and Stannard, Russell. *The New World of Mr. Tompkins*. Cambridge University Press, Cambridge MA, 1999.

Garfinkle, David and Garfinkle, Richard. *Three Steps to the Universe from the Sun to Black Holes to the Mystery of Dark Matter*. The University of Chicago Press, Chicago, 2008.

Gasperini, Maurizio. *The Universe Before the Big Bang Cosmology and String Theory.* Springer, Berlin, Heidelberg, 2008.

Gates, Evalyn. *Einstein's Telescope The Hunt For Dark Matter and Dark Energy in the Universe.* W. W. Norton & Company, New York, 2009.

Geroch, Robert. *General Relativity From A to B.* The University of Chicago Press, Chicago, 1978.

Gould, Stephen Jay. *The Structure of Evolutionary Theory.* The Belknap Press of Harvard University Press, Cambridge MA, 2002.

Graneau, Peter and Graneau, Neal. *In the Grip of the Distant Universe The Science of Inertia.* World Scientific Publishing Co. Pte. Ltd., USA office Hackensack NJ, 2006.

Greene, Brian R. *The Elegant Universe.* W. W. Norton & Company, Inc., New York, 2003.

Greene, Brian R. *The Fabric of the Cosmos.* Alfred A. Knopf, New York, 2004.

Halpern, Paul and Wesson, Paul. *Brave New Universe Illuminating the Darkest Secrets of the Cosmos.* Joseph Henry Press, Washington D.C., 2006.

Halpern, Paul. *Collider.* John Wiley & Sons, Inc., Hoboken NJ, 2009.

Hawking, Stephen W. *A Brief History of Time*. Bantam Books, New York, 1988.

Hester, Jeff, et al. *21ˢᵗ Century Astronomy*. First edition. W.W. Norton & Company, Inc., New York, 2002

Hewitt, Paul G. *Conceptual Physics*. Addison-Wesley Publishing Co., 11ᵗʰ edition, Reading MA, 2010.

Hey, Tony and Walters, Patrick. *The New Quantum Universe*. Cambridge University Press, Cambridge MA, 2003.

Hooper, Dan. *Dark Cosmos In Search of our Universe's Missing Mass and Energy*. Smithsonian Books in association with HarperCollins Publishers, New York, 2006.

Hooper, Dan. *Nature's Blueprint: Supersymmetry and the Search for A Unified Theory of Matter and Force*. HarperCollins Publishers, New York, 2008.

Kennefick, Daniel. *Traveling at the Speed of Thought*. Princeton University Press, Princeton, 2007.

Koestler, Arthur. *The Ghost in the Machine*. Random House, Inc., New York, 1976.

Krane, Kenneth S. *Introductory Nuclear Physics*. John Wiley & Sons, Hoboken NJ, 1988.

Kutner, Marc L. *Astronomy A Physical Perspective*. Cambridge University Press, 2nd edition, Cambridge MA, 2003.

Larson, Edward J. *Evolution: The Remarkable History of a Scientific Theory*. Modern Library an imprint of The Random House Publishing Group, New York, 2004.

Laszlo, Ervin. *The Systems View of the World*. George Braziller, Inc., New York, 1988.

Laughlin, Robert B. *A Different Universe* (*Reinventing Physics from the Bottom Down*). Basic Books, New York, 2005.

Lederman, Leon M. *Symmetry and the Beautiful Universe*. Prometheus Books, Amherst NY, 2004.

Lincoln, Don. *The Quantum Frontier The Large Hadron Collider*. The Johns Hopkins University Press, Baltimore, 2009.

Lindley, David. *Uncertainty Einstein, Heisenberg, Bohr, and the Struggle for the Soul of Science*. Doubleday, New York, 2007.

Moffitt, John W. *Reinventing Gravity A Physicist Goes Beyond Einstein*. HarperCollins Publishers, New York, 2008.

Morrison, Ian. *Introduction to Astronomy and Cosmology*. John Wiley and Sons, Ltd., Chichester UK, 2008.

Munowitz, Michael. *Knowing the Nature of Physical Law*. Oxford University Press, Cambridge MA, 2005.

Overduin, James M. and Wesson, Paul S. *The Light/Dark Universe*. World Scientific Publishing Co. Pte. Ltd., USA office Hackensack NJ, 2008.

Padmanabhan, Thanu. *Quantum Themes The Charms of the Microworld*. World Scientific Publishing Co. Pte. Ltd., USA office Hackensack NJ, 2009.

Pierce, John R. *Almost All About Waves*. MIT Press, 1974. Dover Publications, Inc., Mineola NY, 2006.

Prigogine, Ilya. *From Being to Becoming: Time and Complexity in the Sciences*. Freeman, San Francisco, 1980.

Prigogine, Ilya and Stengers, Isobelle. *Order Out of Chaos*. Heinemann, London, 1984.

Schneider, Eric D. and Sagan, Dorion. *Into the Cool: Energy Flow, Thermodynamics and Life*. The University of Chicago Press, Chicago, 2005.

Sciama, D.W. *Modern Cosmology*. Cambridge University Press, Cambridge MA, 1975.

Sidharth, B G. *The Thermodynamic Universe: Exploring the Limits of Physics*. World Scientific Publishing Co. Pte. Ltd., USA office Hackensack NJ, 2008.

Sklar, Lawrence. *Space, Time, and Spacetime*. University of California Press, Berkeley CA, 1976.

Skyttner, Lars. *General Systems Theory—Problems, Perspectives, Practice*. World Scientific Publishing Co. Pte. Ltd., USA office Hackensack NJ, 2005.

Smolin, Lee. *Three Roads to Quantum Gravity*. Basic Books, New York, 2001.

Smolin, Lee. *The Life of the Cosmos*. Oxford University Press, New York, 1997.

Smolin, Lee. *The Trouble With Physics The Rise of String Theory, The Fall of a Science and What Comes Next*. Houghton Mifflin Company, New York, 2006.

Steinhardt, Paul J. and Turok, Neil. *Endless Universe Beyond the Big Bang*. Doubleday Broadway Publishing Group, New York, 2007.

Strickberger, Monroe W. *Evolution*, Second edition. Jones and Bartlett Publishers, Sudbury MA, 1996.

Styer, Daniel F. *The Strange World of Quantum Mechanics*. Cambridge University Press, Cambridge MA, 2004.

Swenson, R. *Autocatakinetics, Evolution, and the Law of Maximum Entropy Production: A Principled Foundation Toward the Study of Human Ecology*. Advances in Human Ecology, vol. 6, pp. 1-47, 1997.

Swenson, R. *Spontaneous Order, Autocatakinetic Closure, and the Development of Space-Time*. Annals New York Academy of Sciences, vol. 901, pp. 311-319, 2000.

Taylor, John. *Black Holes The End of the Universe?* Souvenir Press Ltd., London, 1973. Guernsey Press Co. Ltd., Guernsey, Channel Islands, 1998.

Tolman, Richard C. *Relativity Thermodynamics and Cosmology*. Oxford University Press, Inc., 1934. Dover Publications, Inc., New York, 1987.

Uzan, Jean-Philippe and Leclercq, Benedicte. *The Natural Laws of the Universe Understanding Fundamental Constants*. Praxis Publishing Ltd., Chichester UK, 2008.

Van Ness, H. C. *Understanding Thermodynamics*. McGraw-Hill Book Company, 1969. Dover Publications Inc., New York, 1983.

Veltman, Martinus. *Facts and Mysteries in Elementary Physics*. World Scientific Publishing Co. Pte. Ltd., USA office Hackensack NJ, 2003.

Wilczek, Frank. *The Lightness of Being: Mass, Ether, and the Unification of Forces*. Basic Books, New York, 2008.

Woit, Peter. *Not Even Wrong the Failure of String Theory and the Search for Unity in Physical Law*. Basic Books, New York, 2006.

www.ingramcontent.com/pod-product-compliance
Lightning Source LLC
Chambersburg PA
CBHW021950170526
45157CB00003B/927